Industrial Water Treatment

Editor

Gyan Singh

scitus
academics

Industrial Water Treatment

Edited by **Gyan Singh**

ISBN: 978-1-68117-406-8

Library of Congress Control Number: 2015941603

SCITUS Academics LLC,
www.scitusacademics.com
Box No. 4766, 616 Corporate Way,
Suite 2, Valley Cottage,
NY 10989

Printed in United States of America on Acid Free Paper ∞

Contents

Preface

Water Treatment Field Manual presents different methods used in produced water treatment systems in the oil and gas industry. Produced water is salty water that is produced as a by-product along with oil or gas during the treatment. Water is brought along with the oil and gas when these are lifted from the surface. The water is then treated before the discharge or re-injection process. Industrial Wastewater Treatment, Recycling and Reuse is an accessible reference to assist you when handling wastewater treatment and recycling. It features an instructive compilation of methodologies, including advanced physicochemical methods and biological methods of treatment. It focuses on recent industry practices and preferences, along with newer methodologies for energy generation through waste. The book discusses the basic terms and concepts that describe produced water treatment. It also presents the different methods involved in the treatment. It further discusses the design, operation, maintenance, and sizing of the produced water treatment systems. In the latter part of the book, the ways to remove impurities in water are discussed, including choosing the proper filter, filtering equipment, filtering methods, and filtering types.

Editor

Chapter 1

A Review on Anaerobic–aerobic Treatment of Industrial and Municipal Wastewater

Yi Jing Chan, Mei Fong Chong, Chung Lim Law, and D.G. Hassell

School of Chemical and Environmental Engineering, Faculty of Engineering, The University of Nottingham Malaysia Campus, Jalan Broga, 43500 Semenyih, Selangor, Malaysia

ABSTRACT

Anaerobic–aerobic systems have been remarkably employed in industrial and municipal wastewater treatment for many years. While previously most treatment of wastewaters have been carried out in conventional anaerobic–aerobic treatment plants, in recent years, high rate anaerobic–aerobic bioreactors have been increasingly employed for wastewaters with high chemical oxygen demand (COD). This paper provides a review of the various types of high rate anaerobic–aerobic water treatment techniques currently available including high rate bioreactors and integrated anaerobic–aerobic bioreactors. The integrated bioreactors are classified into four types, which are (i) integrated bioreactors with physical separation of anaerobic–aerobic zone, (ii) integrated bioreactors without physical separation of anaerobic–aerobic zone, (iii) anaerobic–aerobic Sequencing Batch Reactors (SBR), and (iv) combined anaerobic–aerobic culture system. The integration of aerobic and anaerobic degradation pathways in a single bioreactor is capable of enhancing the overall degradation efficiency. The merits of different integrated anaerobic–aerobic bioreactors are highlighted and comparison made to identify possible future areas of research to fully utilize these methods of wastewater treatment. The comparison demonstrates that using an integrated bioreactor with stacked configuration in treating high strength industrial wastewaters is advantageous due to minimal space requirements, low capital cost and excellent COD removal efficiencies (in excess of 83%).

INTRODUCTION

Over the last century, continued population growth and industrialization have resulted in the degradation of various ecosystems on which human life relies on. In the case of ocean and river quality, such pollution is primarily caused by the discharge of inadequately treated industrial and municipal wastewater. On initial discharge, these wastewaters can contain high levels of inorganic pollutants which can be easily biodegradable, but whose impact load on the ecosystems, either in Total Suspended Solids (TSS), Biochemical Oxygen Demand (BOD_5), or Chemical Oxygen Demand (COD), may be in the tens of thousands mg/L [1]. To combat this increasing burden on our aquatic environment, increasingly strict regulation on pollution discharge is being implemented by various governmental bodies, with focus primarily on waste reduction. The treatment systems developed by industry are frequently regarded as a regulatory obligation, increasing capital and running costs and yielding negative economic returns. Compliance to environmental legislations should not necessary lead to the creation of additional costs, but can instead provide a secondary source of income. One possible source of increased revenue available to industries is through taking advantage of the incentives awarded by the Clean Development Mechanism (CDM) under the Kyoto Protocol 1997.

In the treatment of wastewater, biological treatment appears to be a promising technology to attain revenue from Certified Emission Reduction (CER) credits, more commonly known as carbon credits from the CDM as methane gas is generated from anaerobic digestion and can be utilized as renewable energy. With appropriate analysis and environmental control, almost all wastewaters containing biodegradable constituents with a BOD/COD ratio of 0.5 or greater can be treated easily by biological means [2]. In comparison to other methods of wastewater treatment, it also has the advantages of lower treatment costs with no secondary pollution [3]. Both aerobic and anaerobic processes can be used; the former involves the use of free or dissolved oxygen by microorganisms (aerobes) in the conversion of organic wastes to biomass and CO_2 while in the latter complex organic wastes are degraded into methane, CO_2 and H_2O through three basic steps (hydrolysis, acidogenesis including acetogenesis and methanogenesis) in the absence of oxygen. Aerobic biological processes are commonly used in the treatment of organic wastewaters for achieving high degree of treatment efficiency, while in anaerobic treatment, considerable progress has been achieved in anaerobic biotechnology for waste treatment based on the concept of resource recovery and utilization while still achieving the objective of pollution control [4] and [5].

The various merits of both treatments are highlighted in Table 1, and both systems are capable of achieving high organic removal efficiency. In general, aerobic systems are suitable for the treatment of low strength wastewaters (biodegradable COD concentrations less than 1000 mg/L) while anaerobic systems are suitable for the treatment of high strength wastewaters (biodegradable COD concentrations over 4000 mg/L). According to Cakir and Stenstrom

[6], there exist cross over points, ranging from 300 to 700 mg/L influent wastewater ultimate BOD (BOD_u), which are crucial for effective functioning of aerobic treatment systems. The advantages of anaerobic treatment outweigh the advantages of aerobic treatment when treating influents in higher concentrations than the cross over values, and generally anaerobic treatment requires less energy with potential bioenergy and nutrient recovery. However, compared to anaerobic systems, aerobic systems achieve higher removal of soluble biodegradable organic matter material and the produced biomass is generally well flocculated, resulting in lower effluent suspended solids concentration [7]. As a result, the effluent quality from an aerobic system is generally higher than the anaerobic system.

Table 1: Comparison of aerobic and anaerobic treatment [4] and [7]

Feature	**Aerobic**	**Anaerobic**
Organic removal efficiency	High	High
Effluent quality	Excellent	Moderate to poor
Organic loading rate	Moderate	High
Sludge production	High	Low
Nutrient requirement	High	Low
Alkalinity requirement	Low	High for certain industrial waste
Energy requirement	High	Low to moderate
Temperature sensitivity	Low	High
Startup time	2–4 weeks	2–4 months
Odor	Less opportunity for odors	Potential odor problems
Bioenergy and nutrient recovery	No	Yes
Mode of treatment	Total (depending on feedstock characteristics)	Essentially pretreatment

Highly polluting industrial wastewaters are preferably treated in an anaerobic reactor due to the high level of COD, potential for energy generation and low surplus sludge production. However in practical applications, anaerobic treatment suffers from the low growth rate of the microorganisms, a low settling rate, process instabilities and the need for post treatment of the noxious anaerobic effluent which often contains ammonium ion (NH_4^+) and hydrogen sulfide (HS^-) [8]. In most applications, despite the efficiency of the anaerobic process is high, complete stabilization of the organic matter is impossible anaerobically due to the high organic strength of the wastewater. The final effluent produced by the anaerobic treatment contains solubilized organic matter. This is suitable for aerobic treatment, indicating the potential of using anaerobic–aerobic systems [9] and subsequent post treatment using aerobic treatment is required to meet the effluent discharge standard.

For green olive debittering wastewater with a COD varying between 25,000 and 100,000 mg/L, Aggelis et al. [10] found that neither anaerobic nor aerobic processes could be employed alone for efficient treatment. When treating this high organic strength industrial wastewaters, aerobic or anaerobic treatment alone do not produce effluents that comply with effluent discharge limit. The use of anaerobic–aerobic processes can also lead to a factor eight cost reduction in operating costs when compared with aerobic treatment alone [11], while simultaneously resulting in high organic matter removal efficiency, a smaller amount of aerobic sludge and no pH correction. Benefits of the anaerobic–aerobic process have been identified by Frostell [12] and Cervantes et al. [13] are listed below:

- Great potential of resource recovery: Anaerobic pretreatment removes most of the organic pollutants and converts them into a useful fuel, biogas.
- High overall treatment efficiency: Aerobic post-treatment polishes the anaerobic effluent and results in very high overall treatment efficiency. The aerobic treatment also smoothes out fluctuations in the quality of the anaerobic effluent.
- Less disposal of sludge: By digesting excess aerobic sludge in the anaerobic tank, a minimum stabilized total sludge is produced which leads to a reduction in sludge disposal cost. As an additional benefit, a higher gas yield is achieved.

- Low energy consumption: anaerobic pretreatment acts as an influent equalization tank, reducing diurnal variations of the oxygen demand and resulting in a further reduction of the required maximum aeration capacity.
- When volatile organics are present in the wastewater, the volatile compound is degraded in the anaerobic treatment, removing the possibility of volatilization in the aerobic treatment.

Thus it can be seen that it is operationally and economically advantageous to adopt anaerobic–aerobic processes in the treatment of high strength industrial wastewaters since it couples the benefit of anaerobic digestion (i.e. biogas production) with the benefits of aerobic digestion (i.e. better COD and volatile suspended solid (VSS) removal) [14]. As well as their capability to biodegrade organic matter, anaerobic–aerobic systems have also been found to perform well for the following processes: biodegradation of chlorinated aromatic hydrocarbons including anaerobic dechlorinations and aerobic ring cleavage [15]; sequential nitrogen removal including aerobic nitrification and anaerobic denitrification [16]; anaerobic reduction of Fe(III) and microacrophilic oxidation of Fe(II) with production of fine particles of iron hydroxide for adsorption of organic acids, phenols ammonium, cyanide, radionuclides, and heavy metals [17].

These advantages have prompted the rapid development of anaerobic–aerobic systems in the treatment of both industrial wastewater [18], [19], [20], [21] and [22] and municipal wastewater (primarily designed for nutrient removal) [23], [24] and [25]. While most treatment of industrial and municipal wastewaters has been carried out in conventional anaerobic–aerobic treatment plants, high rate bioreactors have been developed to reduce the capital cost of the process. However, the investigation on the high rate anaerobic–aerobic treatment are limited to a few studies and not well documented. Hence, this review aims to summarize and discuss the feasibility of high rate anaerobic–aerobic treatment for efficient organic removal of industrial and municipal wastewater.

This review also provides an overview of different types of anaerobic–aerobic treatment system, providing a comparison between the conventional and newer technologies.

TYPES OF ANAEROBIC–AEROBIC TREATMENT SYSTEMS

Fig. 1 highlights the three main types of anaerobic–aerobic system currently in use, with distinctions made between the different approaches used to obtain an anaerobic–aerobic reactor system.

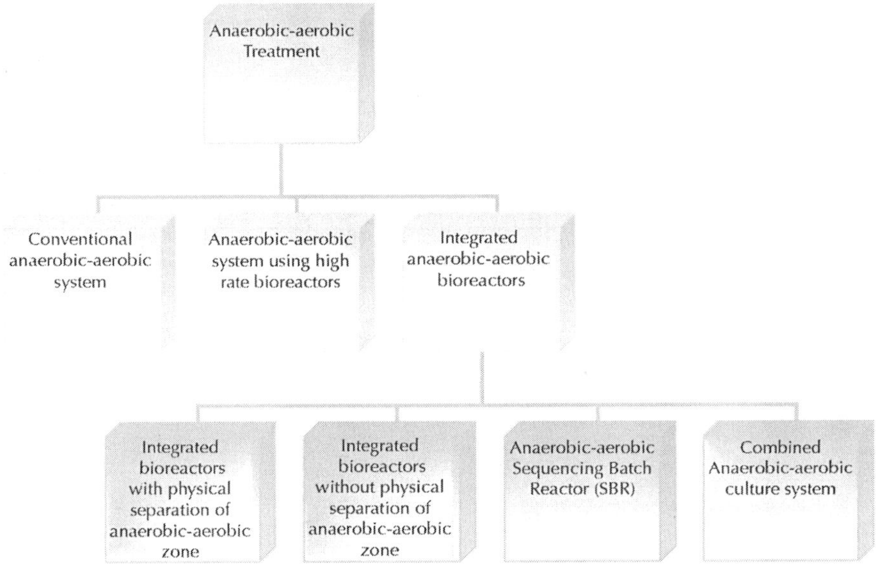

Figure 1: Types of combined anaerobic–aerobic system.

The simplest approach for the anaerobic–aerobic treatment is the use of conventional systems such as aerated stabilization ponds, aerated and non-aerated lagoons, as well as natural and artificial wetland systems. Aerobic treatment occurs in the upper part of these systems while anaerobic treatment occurs at the bottom end.

A typical organic loading is 0.01 kg BOD/m^3 day and the retention time varies from a few days to 100 days [17].

Conventional anaerobic–aerobic systems usually comprise large ponds connected in series and are frequently characterized by long hydraulic retention time (HRT), low organic loading rate (OLR), as well as vast area of land or digesters. In short, the conventional treatment plants suffer from problems related to their large space requirement, emissions into populated environments from large open reactors, low process efficiencies, large surplus sludge production and high energy consumption. These eventually decrease the attractiveness of conventional anaerobic–aerobic treatment plants for reasons of economy and location.

New technologies have been developed over the years to overcome the disadvantages of conventional anaerobic–aerobic systems. Anaerobic–aerobic system using high rate bioreactors (such as upflow anaerobic sludge blanket (UASB), filter bioreactor, fluidized bed reactor, membrane bioreactor) are adopted in order to provide a treatment process which is both technologically and economically viable with the dual goals of resource recovery and compliance with current legislation for effluent discharge. A more intensive form of biodegradation can also be achieved by integrating anaerobic and aerobic zones within a single bioreactor. Essentially, there are four types of integrated anaerobic–aerobic bioreactor. These are (i) integrated bioreactors with physical separation of anaerobic–aerobic zone, (ii) integrated bioreactors without physical separation of anaerobic–aerobic zone, (iii) Sequencing Batch Reactors (SBR) based on temporal separation of the anaerobic and the aerobic phase, and (iv) combined anaerobic–aerobic culture system based on the principle of limited oxygen diffusion in microbial biofilms. An overview of the most frequently applied bioreactors is delineated in Table 2, Table 3 and Table 4, with specific attention on the evaluation of their treatment efficiency in terms of organic removal.

Table 2: Anaerobic–aerobic systems using high rate bioreactors

No	Type[a]	Type of wastewater[b]	Influent COD (mg/L)	OLR[c] (kg COD/ m³ d)	Total COD removal (%)	Anaerobic COD removal (%)	Aerobic COD removal (%)	Total HRT[d] (h or d)	Anaerobic HRT[d] (h or d)	HRT[d](h or d)	Reference
1	UASB + CSTR	Wool acid dying ww	499–2000	–	83–97	51–84	–	3.3 d	17 h	–	[26]
2	UASB + CSTR	Cotton textile mill ww	604–1038	–	40–85	9–51	–	5.75 d	30 h	4.5 d	[27]
3	UASB + CSTR	Simulated textile ww	4214	1.01–15.84	91–97	–	–	19.17–1.22 d	–	–	[28]
4	2 UASBs + CSTR	Food solid waste leachate	5400–20000	4.3–16	96–98	58–79	85–89	5.75 d	1.25 d	4.5 d	[20]
5	UASB + CSTR	Pulp and paper industry effluent	5500–6600	16	91	85	–	11.54 h	5 h	6.54 h	[29]
6	UASB + CSTR	Pharmaceutical industry ww	3000	3.6	97	68–89	71–85	–	–	–	[30]
7	UASB + AS	Olive mill ww + municipal ww	1800–4400	3–7	95–96	70–90	>60	28.3 h	14.7 h	13.6 h	[31]
8	UASB + AS	Starch Industry ww	20000	15	–	77–93	64	5 d	1 d	4 d	[32]
9	UASB + AS	Municipal ww	386–958	–	85–93	69–84	43–56	6.8 h	4 h	2.8 h	[33]
10	UASB + AFB	Synthetic textile ww	2000–3000	–	–	–	–	2.7–32.7 h	1.4–20 h	1.3–12.7 h	[34]
11	UASB + AFB	Synthetic textile ww	2700	4.8	80	50	60	20 h	10 h	10 h	[35]
12	RBC + SBR	Mixture of cheese whey and dairy manure	37400–65700	5.2–14.1	99	46.3–62.6	93–95	–	2–5 d	–	[21]

13	RBC + SBR	Screened dairy manure	39900–40100	8.2–26.8	98	18.7–29	86–87	–	1–4 d	–	[36]
14	FFB + FFB	Slaughter house ww	400–1600	0.39	92	–	–	4.7–7.3 d	1.2 d	3.5–6.1 d	[37]
15	EGSB + Aerobic biofilm reactor	POME	35000	10	95.6	93	22	–	3 d	–	[22]
16	UBF + MBR	Synthetic ww	6000–14500	7.2	99	98	–	1 d	–	–	[18]
17	UASB + Aerobic solid contact system	Municipal ww	341	2.6	–	34	–	3.53–6.2 h	3.2 h	0.33–3 h	[23]
18	UASB + RBC	Domestic sewage	640	–	84–95	35–47	52–56	6–13.5 h	3–6 h	3–7.5 h	[24]
19	CSTR + Activated sludge	Green olive debittering ww	23500	0.47	83.5	37.4–48.9	73.6	55 d	50 d	5 d	[10]
20	AFFFBR + AS	PTA effluent	5000	4–5.0	96.4	64–62	90	23–27.2 h	1–1.2 h	22–26 h	[38]
21	AnFB + Air lift suspension reactor	Complex industrial ww	3800	25–30	–	60–65	–	3.4–4.3 h	1.4–1.8 h	2–2.5 h	[8]
22	Hybrid bioreactor + AS	Oil shale ash dump leachate	2000–4600	–	75	20–40	60–80	8.8 d	62 h	150 h	[39]
23	Packed column reactor + AS	Textile industry ww	800–1200	–	50–85	30–65	40–90	22–82 h	12–72 h	10 h	[40]

[a]Reactor Type: UASB, upflow anaerobic sludge bed; CSTR, continuously stirred tank reactor; AS, activated sludge; AFB, aerobic fluidized bed; RBC, rotating biological contactors; SBR, sequencing batch reactor; FFB, fixed film bioreactor; EGSB, expanded granular sludge bed; UBF, upflow bed filter; AFFFBR, anaerobic fixed film fixed bed reactor; AnFB, anaerobic fluidized bed.

[b]Type of waste water (ww): POME, palm oil mill effluent; PTA, purified terephthalic acid.

[c]OLR, Organic loading rate.

[d]HRT, Hydraulic retention time: h, hour; d, day.

Table 3: Integrated bioreactors with physical separation of anaerobic and aerobic zones

No	Bioreactor Type[a]	Type of Wastewater[b]	Influent COD (mg/L)	OLR[c] (kg COD/ m^3 d)	Total COD removal (%)	Anaerobic COD removal (%)	Aerobic COD removal (%)	Total HRT[d](h)	Reference
1	Bubble column with a draught tube	Synthetic ww	–	–	–	–	–	3–11	[69]
2	RAAIB bioreactor	Sewage	345	–	84	–	–	1.2–15.5	[68]
3	SAA bioreactor	Diluted landfill leachate	1000–3300	–	85–95	–	–	–	[70]
4	Anaerobic aerobic integrative baffled reactor	Potato starch ww	1100–4500	–	88.4–98.7	87	–	6–24.0	[67]

[a]Bioreactor Type: RAAIB, Radial anaerobic/aerobic immobilized biomass; SAA, Simultaneous aerobic and anaerobic bioreactor.

[b]Type of wastewater (ww).

[c]OLR, Organic loading rate.

[d]HRT, Hydraulic retention time: h, hour.

Table 4: Integrated bioreactors without physical separation of anaerobic and aerobic zones

No.	Type	Type of Wastewater[a]	Influent COD (mg/L)	OLR[b] (kg COD/ m^3 d)	Total COD removal (%)	Anaerobic COD removal (%)	Aerobic COD removal (%)	Total HRT[c](h or d)	Reference
1	Upflow anaerobic/aerobic fixed bed integrated reactor	Synthetic ww	365–3500	0.8–7.6	95–98	27–70	37–92	9 h	[84]

2	Anaerobic–aerobic granular biofilm reactor	Degradation of Aroclor 1242	–	–	–	–	–	2.1 d	[82]
3	Anaerobic–aerobic granular biofilm reactor	Degradation of TCE	800	–	–	–	–	17–20 h	[66]
4	Methanogenic–methanotrophic hybrid reactor	Degradation of PCE & TCE	–	–	–	–	–	0.5–3 d	[78]
5	Anaerobic–aerobic granular biofilm reactor	Synthetic ww	–	2.89–3.75	95–98	62–95	0–33	48 h	[79]
6	Staged anaerobic–aerobic membrane bioreactor	Synthetic ww	1300–10500	10.08	>99	60–80	–	–	[83]
7	Integrated anaerobic–aerobic fixed film reactor	Slaughter-house ww	1190–2800	0.77	93	0.6–1.2	97	0.94–3.8 d	[81]
8	Integrated anaerobic–aerobic fluidized bed reactor	Municipal ww	350	<1.2	>80	–	–	24 h	[80]

[a]Type of wastewater (ww): TCE, Trichloroethylene; PCE, tetrachloroethelene.

[b]OLR, Organic loading rate.

[c]HRT, Hydraulic retention time: h, hour; d, day.

ANAEROBIC–AEROBIC SYSTEMS USING HIGH RATE BIOREACTORS

The right combination and sequence of treatment methods is the key to the successful handling of industrial and municipal wastewater. The combinations of different anaerobic and aerobic bioreactors have been applied to treat a broad range of industrial wastewater including textile industry wastewater, food solid waste leachate, pulp and paper industry wastewater, pharmaceutical industry wastewater, mixture of olive oil mill wastewater and primary municipal wastewater, starch industry wastewater, green olive debittering wastewater, slaughter house wastewater, and palm oil mill effluent (POME). Table 2 lists the anaerobic–aerobic systems using high rate bioreactors where the treatment is carried out in two separate bioreactors connected in series. The anaerobic–aerobic systems using high rate bioreactors reviewed in this paper achieve high COD removal (in excess of 70%) at short HRT (ranging from few hours to few days). Therefore, the anaerobic–aerobic treatment is an efficient method to treat industrial and municipal wastewater.

Upflow Anaerobic Sludge Bed (UASB) and Continuous Stirred Tank Reactor (CSTR) System

Upflow Anaerobic Sludge Bed (UASB) reactors appear to be a robust technology and have performed well in wastewater treatment for a number of decades [5]. They are extensively exploited as a pre-treatment before aerobic treatment for various types of industrial and municipal wastewaters. It has shown to be a technology proficient in overcoming some of the disadvantages of the mechanized aerobic systems, due to the lower energy consumption and sludge generation. A schematic of an UASB system is shown in Fig. 2. It consists of an upflow of wastewater through a dense sludge bed with high microbial activity [41] and [42]. The processes are based on the development of dense granules (with diameter of 1–4 mm) formed by the natural self-immobilization of the anaerobic bacteria, considered to be an essential condition for the successful operation of a UASB reactor. Suspended and colloidal components of wastewaters in the form of fat, protein, and cellulose have adverse impact on the performances of UASB reactors, preventing the system from operating at high organic loading rates and can cause deterioration of microbial activities and wash out of active biomass [43]. Generally, UASB are reported to remove more than 60% of COD from most types of wastewater. Impressive though this is, UASB treated effluent usually fails to comply with most of discharge standards [33].

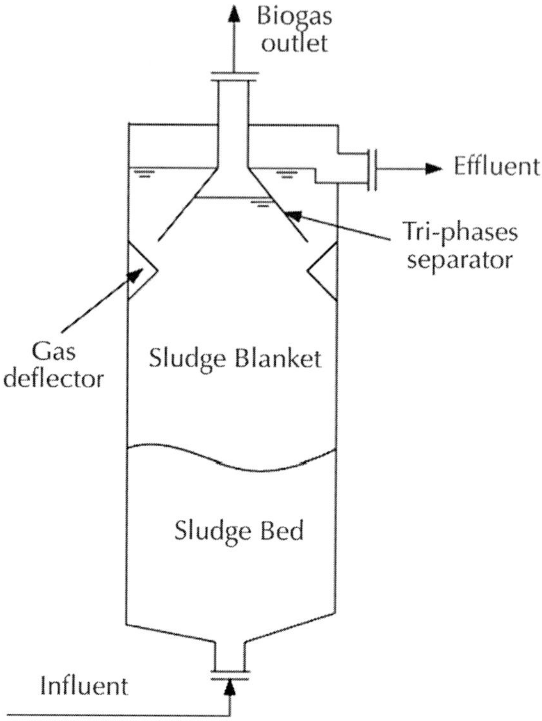

Figure 2: Schematic representation of an UASB reactor [41]. (Reprinted from Biological Wastewater Treatment in Warm Climate Regions, p. 723, ISBN 9781843390022 with permission from the copyright holders, IWA).

Activated sludge, which consists of stirred and aerated flocculated suspension of a mixed bacterial population that comes into contact with wastewater, is the most commonly used process in aerobic treatment [44]. It has high efficiency with operational flexibility and possibility for nutrient removal. The activated sludge process can be a plug flow reactor (PFR), continuously stirred tank reactor (CSTR), or sequencing batch reactor (SBR). However, operating activated sludge system alone requires a high level of mechanization, high construction and operational costs, sophisticated operation and the need for treating a substantial amount of sludge.

Balancing the advantages and disadvantages of both systems, a combined technology, consisting of an UASB reactor for anaerobic pretreatment, followed by activated sludge for aerobic post treatment has been extensively employed [32] and [33]. A significant feature of this system is the return of the excess aerobic sludge to the UASB reactor where the solids undergo stabilization, and thus simplify the sludge treatment. The overall sludge production of the anaerobic–aerobic system is wasted only from the UASB reactor. Since it is already thickened and stabilized, it can be directly sent for dewatering and final disposal.

Several authors have validated the utility of UASB/aerobic CSTR system for tackling a wide variety of industrial wastewaters with BOD/COD ratio ranging from 0.17–0.74 [20], [26], [27], [28], [29], [30], [31], [32] and [33]. Based on the data shown in Table 2, the UASB/aerobic CSTR systems can typically remove 83–98% of COD with influent COD content in the range of 500–20,000 mg/L at a total HRT of 11.54 h to 6 days.

Lerner et al. [45] presented a comparison between the results of paper mill wastewater treatment of a full-scale activated sludge treatment system working as the only biotreatment and the activated sludge treatment with anaerobic pretreatment by UASB. The anaerobic–aerobic system manifests steady operation performance, while the activated sludge treatment system without anaerobic pretreatment produced effluent with oscillatory quality. The results indicate a higher level of organic matter removal in the anaerobic–aerobic treatment system, with a final discharge of 80–120 mg COD/L as compared to 220–250 mg COD/L for the activated sludge treated effluent. It also illustrated that the anaerobic–aerobic treatment greatly reduced electrical consumption in the biological treatment plant and thus lowers operational costs.

UASB and Aerobic Fluidized Bed (AFB) System

Fluidized bed reactors are packed with mobile supports in which particles covered with biofilm are fluidized by the recirculation of liquid. They eliminate substrate diffusion limitations, which are usually inherent in stationary bed process. A schematic of an aerobic fluidized bed (AFB) system is illustrated in Fig. 3. The AFB reactor exhibits numerous advantages such as a high biomass concentration, high OLR, short HRT, no bed clogging, small external mass transfer resistance and large surface area for mass transfer [46],[47] and [48]. Conversely, there are some problems which inhibit their applicability on a large industrial scale such as control of the bed expansion, thickness of the biofilm and oxygen distribution system as well as high-energy consumption due to the very high liquid recirculation ratio [49] and [50]. The most common operational mode of an AFB reactor in wastewater treatment contains three phases: (i) the discrete solid phase of inert particles with immobilized microbial cells, (ii) the discrete air bubbles and (iii) the continuous aqueous solution.

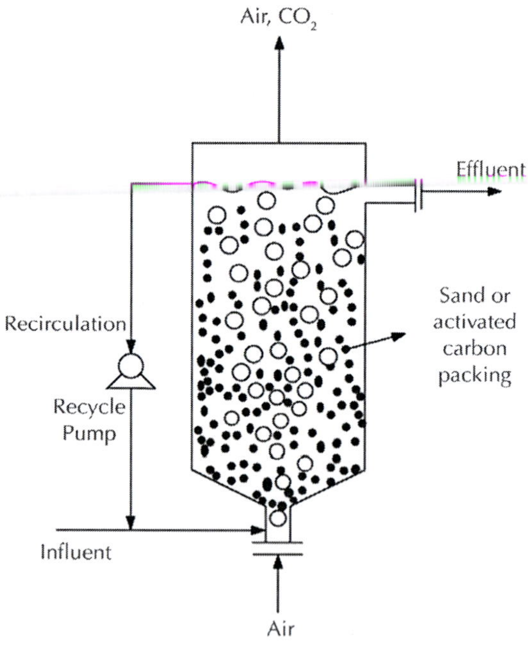

Figure 3: Schematic representation of an AFB reactor [41]. (Modified from Biological Wastewater Treatment in Warm Climate Regions, p. 717, ISBN 9781843390022 with permission from the copyright holders, IWA).

Tavares et al. [51] showed that a high average COD removal efficiency (82%) was obtained in the treatment of a synthetic wastewater with feed COD content of 180 mg/L when the three-phase AFB is operated at a low average HRT (30 min). This result indicates the potential of this reactor to treat low strength wastewaters with COD content in the range of 100–200 mg/L.

By incorporating an UASB with the AFB reactor, a total COD removal efficiency of 75% at an overall HRT of 14 h can be accomplished in the treatment of a medium strength synthetic textile wastewater with 2700 mg COD/L [35]. The results revealed that the combined UASB–AFB system produced 45% lower sludge volume than the aerobic system.

However, Yu et al. [35] pointed out that the anaerobic biomass (~1 g volatile solid (VS)/L) brought into AFB reactor contributed to an increase in suspended solids, rather than improved COD removal, because of its fast deactivation under aerobic conditions. The dead anaerobic cells diluted the specific activity of aerobes. Thus, the cell mass brought from the UASB reactor to the AFB reactor should be minimized to avoid a high turbidity with nil contribution in biological activity by the anaerobes.

The UASB-AFB system is useful in the biological treatment of medium strength industrial wastewaters due to its high pH tolerance, reduced sludge formation and stable COD removal performance. The UASB-AFB configuration emerges as an attractive alternative from technical, economical and environmental points of view, especially when space is a limiting factor.

Anaerobic Rotating Biological Contactors (RBC) and Aerobic Sequencing Batch Reactor (SBR) System

In a rotating biological contactor (RBC) system, microorganisms attach to an inert support medium and form a biological film. The support medium, with a sequential disc configuration, is partly or totally submerged and rotates slowly around a horizontal axis in a tank through which the wastewater flows. A schematic of an RBC reactor is shown in Fig. 4. The system configuration of anaerobic RBC is similar to that of the aerobic RBC, except that the tank is covered to avoid contact with air [41]. When employing anaerobic RBC alone for treating high-strength synthetic wastewater with COD concentrations between 3248 and 12150 mg/L, the final COD of the RBC effluent is still considered too high. Thus, a further treatment is required albeit satisfactory efficiencies of overall COD removal ranged from 74 to 82% are achieved at a HRT of 32 h [52].

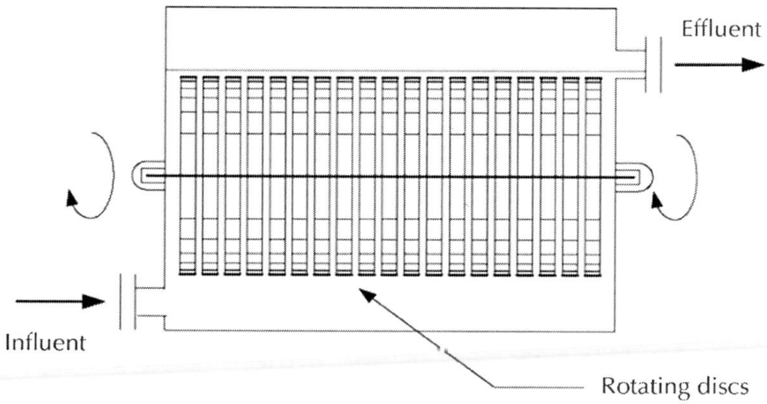

Figure 4: Schematic representation of an anaerobic RBC reactor [41]. (Reprinted from Biological Wastewater Treatment in Warm Climate Regions, p. 719, ISBN 9781843390022 with permission from the copyright holders, IWA).

Advantages of the RBC system are low energy requirements, short retention time, excellent process control, low operating costs and that it is capable of handling a wide range of flows. Disadvantages include its process performance being susceptible to wastewater characteristics, resulting in limited operational flexibility to varying loading and operating conditions [7] and frequent maintenance on its shaft bearings and mechanical drive units.

The aerobic sequencing batch reactor (SBR) is an improved version of the fill and draw activated sludge system, consisting of one or more tanks, each capable of waste stabilization and solids separation [2]. The SBR process offers flexibility in the treatment of variable flows, minimum operator interaction, option for aerobic or anaerobic conditions in the same tank, good oxygen contact with microorganisms and substrate, small floor space, and good removal efficiency [53]. These advantages justify the recent increase in the implementation of this process in industrial [21], [36] and [54] and municipal [55], [56] and [57] wastewater treatment.

Since the anaerobic RBC can attain high methane production rates and the aerobic SBR process can treat dilute wastes efficiently, combining the two processes result in an efficient bioenergy production and waste treatment system. Hence, an anaerobic RBC integrated with three aerobic SBRs has been adopted in the treatment of screened dairy manure and a mixture of cheese whey with dairy manure [21] and [36]. Generally, this combined system is able to achieve apparent COD reduction of at least 98% and also produces substantial amounts of methane gas.

Anaerobic–aerobic Fixed Film Bioreactor (FFB) System

Immobilized cells on the surface (fixed-film) of the media offer some advantages over cultures in suspension such as; a greater variation in population; less sensitivity to environmental variations (temperature, pH, and toxic substances); higher growth rate; faster utilization of the substrate in relation to free biomass. This is attributed to physiological modification of the fixed cells undergo, due to either the increase in the local concentration of nutrients and enzymes, or the selective effect of the extracellular polymeric matrix in relation to inhibitory or toxic substances [58].

The combination of two fixed-film bioreactors (FFB) with arranged media, the first anaerobic and the second aerobic, connected in series with recirculation for treatment of poultry slaughterhouse wastewater was evaluated by Del Pozo and Diez [37]. Owing to the high oil and grease (O&G) content in slaughterhouse wastewater which will cause serious flotation problems in suspended biomass systems, FFB was chosen in this system. Long corrugated PVC tubes were placed vertically as support media to avoid clogging, while the rough structure of the tubes increased their specific surface and protected the attached biomass from stress forces. Promising results were obtained, with an overall COD removal efficiency of 92% at OLR of 0.39 kg COD/m^3 day. A schematic of anaerobic–aerobic FFB is shown in Fig. 5.

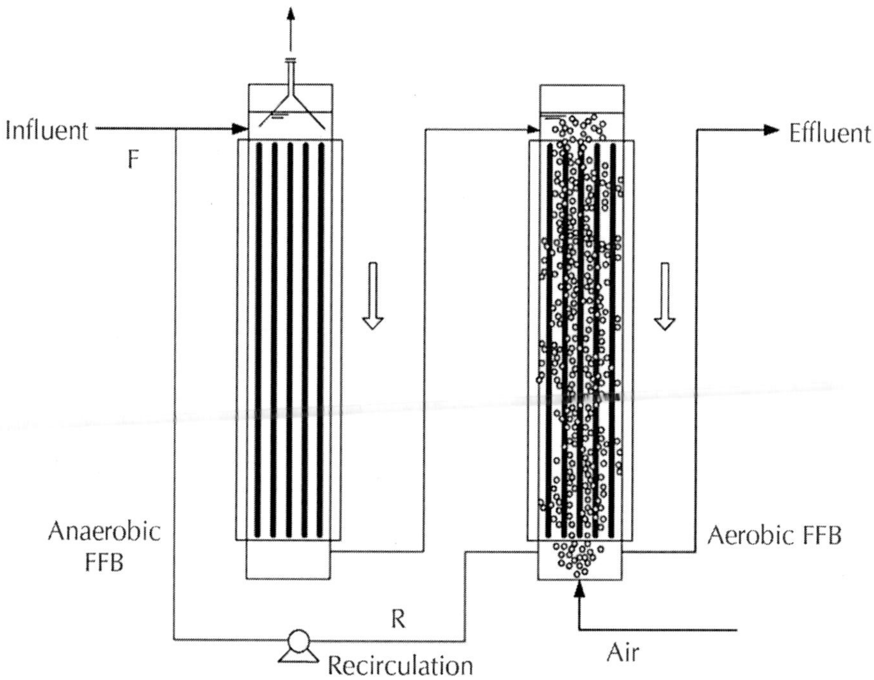

Figure 5: Schematic diagram of anaerobic–aerobic FFBs [37]. (Reprinted from Water Research, Organic matter removal in combined anaerobic-aerobic fixed-film bioreactors, 37 (2003) 3561–3568, with permission from the copyright holders, Elsevier).

Effects of recirculation ratio (R/F) and anaerobic/aerobic volume ratio (V_{an}:V_{ae}) on the fraction of COD removed by each reactor were evaluated. The FFBs were operated in a down flow manner and the aerobic effluent was recirculated to the anaerobic FFB. COD removal occurred mainly in the anaerobic FFB and this effect was accentuated when the recirculation ratio rose from 1 to 6 as a result of the increased contribution of denitrification. Besides, the fraction of COD removed in the anaerobic FFB increased when the volume of the aerobic FFB became smaller than anaerobic FFB. High recirculation in the anaerobic FFB feed favored the denitrification to the detriment of the methanogenic process and the production of biogas.

Expanded Granular Sludge Bed (EGSB) and Aerobic Biofilm Reactor System

The expanded granular sludge bed (EGSB) reactor comes under the family of UASB reactors. It has been primarily developed to improve the substrate–biomass contact within the treatment system by means of expanding the sludge bed with a high upflow liquid velocity (>4 m/h) which intensifies hydraulic mixing and results in better performance and stability than the UASB. The high upflow liquid velocity in the reactor is achieved through the application of a high effluent recirculation rate, coupled with a high height/diameter ratio of around 20 or more [41]. Fig. 6 depicts the schematic diagram of EGSB. They have been successfully applied to treat various kinds of wastewater including brewery wastewater, starch wastewater, molasses alcohol slops, slaughterhouse wastewater, POME, domestic and municipal wastewater [5] and [19]. However, EGSB is not suitable for the removal of particulate organic matter due to the high upflow liquid velocity. The influent suspended solids that are not retained by the granular bed will eventually leave the reactor together with the effluent [59].

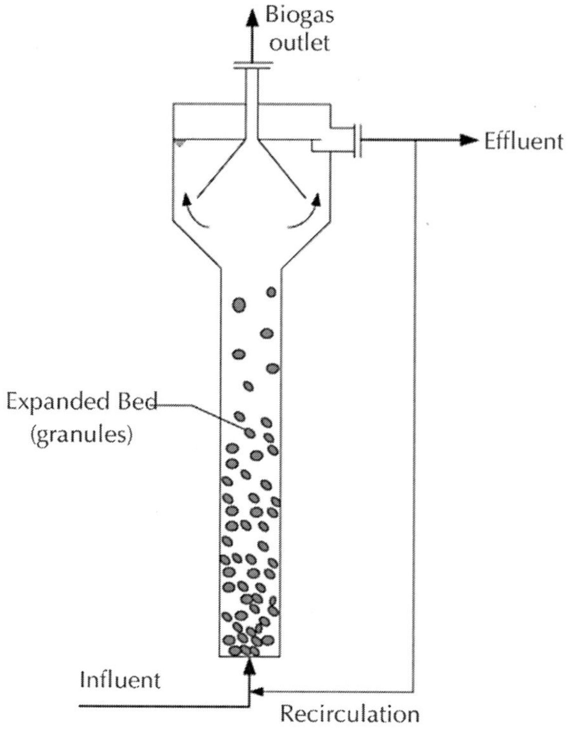

Figure 6: Schematic diagram of EGSB [41].(Reprinted from Biological Wastewater Treatment in Warm Climate Regions, page 724, ISBN 9781843390022 with permission from the copyright holders, IWA).

Through utilizing an EGSB reactor, Zhang et al. [19] reported a high COD removal of 91% for a HRT of 48 h in the treatment of high organic strength wastewater with feed COD content of 80,000 mg/L. However, like other anaerobic systems, the EGSB is still unable to produce final effluent that complies with discharge standards.

In a recent study by Zhang et al. [22] for POME treatment, a total COD reduction of 95.6% at high OLR of 10 kg COD/ (m^3 day) was achieved in a pilot-scale plant composed of an EGSB reactor and aerobic biofilm reactor. The anaerobic EGSB degraded a large portion of organic matter in POME with 93% COD removal while the aerobic biofilm reactor broke down the remaining organic matter (22% of COD removal). In this case the reported average rate of organic matter transformed into methane in the EGSB was only 43% (based on the data of biogas measured), although this could be attributed to the high suspended solids and oil in POME.

Anaerobic Upflow Bed Filter (UBF) and Aerobic Membrane Bioreactor (MBR) System

Anaerobic upflow bed filter (UBF) is an anaerobic hybrid reactor which combines a UASB and anaerobic FFB. The lower part of the UBF reactor is the UASB, where granular sludge is developed. With the presence of stationary packing material, the upper part of the UBF serves as a FFB. The main advantage of the UBF is its ability to eliminate the problems of clogging and biomass washout which are commonly encountered in anaerobic FFB's and UASB's respectively.

Aerobic membrane bioreactors (MBR) combine membrane filtration with biodegradation processes, where solid–liquid separation occurs through sieving. In a MBR, solid materials, biomass, pathogenic bacteria, and even macromolecules are retained while allowing water and smaller solution species to pass through the membrane [60] and [61] so that a very high quality effluent is attained. MBRs offer numerous advantages which include the high quality of the effluent, the separation of solid retention time (SRT) from HRT, the reduced sludge production due to endogenous respiration in long SRT, and low sludge loading rate [62], [63] and [64]. The membrane-retained aqueous and particulate based enzymes which are otherwise lost in the conventional sedimentation–clarification step are also able to improve the metabolic rate in the MBR [65]. While there are numerous advantages, one of the major obstacles in the application of MBR's is membrane fouling, with cross-flow filtration being most commonly used to alleviate this problem.

Ahn et al. [18] reported an apparent COD removal of 99% in the treatment of high-strength wastewater with COD content in the range of 6000–14,500 mg/L by the use of an anaerobic

UBF-aerobic MBR system at a relatively short HRT of 24 h. A schematic diagram of the system is shown in Fig. 7. While this is impressive, it was noted that membrane fouling was observed and the transmembrane pressure (TMP) was about 9 times higher than that of a unit MBR operated under the same conditions. In this case the severe fouling in the system is caused by increased extracellular polymeric substance (EPS) and hydrophobicity.

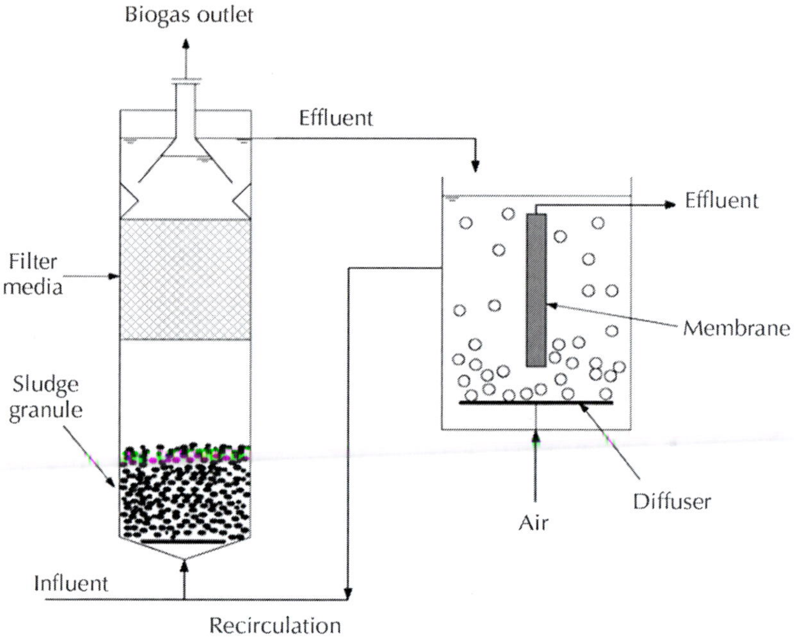

Figure 7: Schematic diagram of UBF-aerobic MBR system [18]. (Reprinted from Desalination, Simultaneous high-strength organic and nitrogen removal with combined anaerobic upflow bed filter and aerobic membrane bioreactor, 202 (2007) 114–121, with permission from the copyright holders, Elsevier).

Process Control and Optimization for COD Removal in Anaerobic–aerobic Systems Using High Rate Bioreactors

Optimization of the COD removal in anaerobic or aerobic bioreactors is more complex than in single stage as the performance of both stages is coupled. To obtain a large degree of depuration at the lowest capital and operational costs, the ideal size of each part of the anaerobic–aerobic system must be determined by an optimization of the anaerobic–aerobic treatment system. It is important to minimize the carbon removed by oxidation in order to reduce oxygen consumption, sludge production and the competition for oxygen between nitrifying and heterotrophic bacteria.

With increased removal of organic matter in the anaerobic reactor, the volume required for the aerobic reactor is reduced with a resulting drop in aeration cost. As aeration is one of the main individual operating costs, optimization of the conversion in the anaerobic reactor and minimization of the aerobic reactor size are exceptionally important. While the anaerobic reactor capital cost increases with increasing anaerobic yields, this increase is rather low compared to the total costs. Minimization of the aerobic reactor size can be carried out with adequate control of a sufficient amount of oxygen so as to maximize the conversion rates. It can be accomplished by utilizing oxygen instead of air which has been proved to reduce the treatment cost significantly [11].

However, if high extent of COD removal is accomplished in the anaerobic reactor, inadequate COD or other nutrients left in the effluent may not favor the performance of aerobes in the following aerobic reactor. Hence, in the optimization of the COD removal in the anaerobic reactor, it is crucial to ensure sufficient but not excessive COD left in the anaerobic bioreactor effluent for effective functioning of the aerobic bioreactor.

The minimization of anaerobic biomass brought into the aerobic reactor is crucial in the optimization of the anaerobic–

aerobic system. As the aerobic reactor accepts an effluent directly from an active anaerobic digester, a significant amount of obligate anaerobes as well as facultative microorganisms enter the aerobic reactor and are not quickly adapted to the aerobic conditions. These active anaerobes could affect the cell population in the aerobic reactor and lead to a mixed microbial population of low oxygen utilization and biological activity. The anaerobic cells will not contribute to COD removal in the aerobic reactor, but increase the suspended solid concentration or the turbidity in the final effluent which puts an extra burden on the downstream sedimentation.

INTEGRATED ANAEROBIC–AEROBIC BIOREACTORS

In recent years, substantial attention has been paid towards the compact high-rate bioreactors for wastewater treatment to meet the strict constraints with respect to space, odor, view, and biosolids production. Thus, the integrated bioreactors which combine the aerobic and anaerobic processes in a single reactor are seen as a viable alternative.

A combination of aerobic and anaerobic degradation pathways in a single reactor is capable of enhancing the overall degradation efficiency [66]. The integrated bioreactors are cost effective, efficient and have smaller foot prints as compared to the aforementioned anaerobic–aerobic systems. Nonetheless, the design, operation and process development of integrated anaerobic–aerobic bioreactors are still in its infancy and limited to a few studies.

Integrated Bioreactors with Physical Separation of Anaerobic and Aerobic Zones

Some approaches have been attempted to obtain the integrated bioreactors by combining anaerobic and aerobic processes in separate zones, such as the baffled reactors [67], the radial anaerobic/aerobic immobilized biomass (RAAIB) reactor [68] and employing an air-lift system for recirculation [69] and [70]. The data pertaining to the performance of those integrated bioreactors with physical separation of the anaerobic and aerobic zones are presented in Table 3.

Bubble Column with a Draught Tube

A cylindrical bubble column with a draught tube as shown in Fig. 8 was employed as a small and simple treatment unit of anaerobic–aerobic activated sludge processes by Hano et al. [69] and more recently Bando et al. [71]. The performances of the unit proposed were evaluated based on the nitrogen removal.

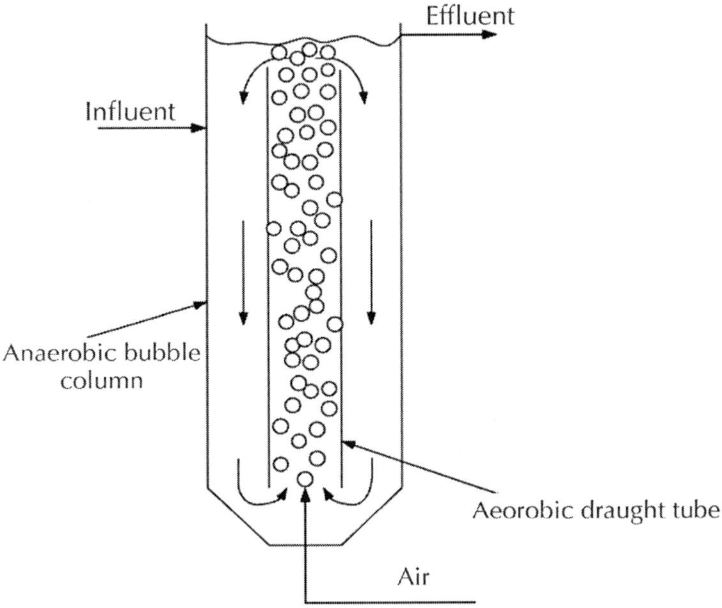

Figure 8: Schematic diagram of the bubble column with a draught tube [69].(Reprinted from Chemical Engineering Science, Biological nitrogen removal in a bubble column with a draught tube, 47 (1992) 3737–3744, with permission from the copyright holders, Elsevier).

The inside of a draught tube is used as an aerobic zone and the annulus as an anaerobic zone. The wastewater is initially introduced to the upper part of the annulus (anaerobic zone) and then flows with the sludge into a draught tube (aerobic zone) by the air-lift action. Finally the effluent is withdrawn from the top of the draught tube.

The volume ratio of anaerobic and aerobic zones in the bubble column can be adjusted simply by changing the diameter of the draught tube. The circulation flow rate of mixed liquor between the two zones is the most important parameter of operation in the anaerobic–aerobic activated sludge processes. It can be varied by changing the height of the draught tube and the flow rate of air into the draught tube. The increase in the circulation rate will cause aerobic conditions to prevail in the annulus which is meant to be anaerobic. As a result, the bubble-column treatment unit should be operated at the lowest circulation rate that can keep the sludge in suspending state.

The advantage associated with the bubble column is no requirement for additional equipment to circulate the mixed liquor between aerobic and anaerobic compartments. Hano et al. [69] demonstrated that the bubble column can be used as a small scale treatment unit, since a satisfactory performance is achieved with relatively simple apparatus and operation. However, the residence time in each zone during the circulation of liquid is too short compared to other anaerobic–aerobic activated sludge processes. Longer residence times are difficult to obtain as a minimum circulation flow rate is required to keep the sludge in a suspended state.

There are several studies which employed rectangular airlift bubble columns installed with support material for enhanced nitrogen removal [72], [73] and [74]. The installation of the support material reduces the minimum circulation flow rate. The anaerobic and aerobic regions are separated by using a partitioning plate in the rectangular airlift bubble column rather than a draft tube, as the partitioning plate is more easily inserted into the existing tank.

Radial Anaerobic–aerobic Immobilized Biomass (RAAIB) Reactor

Immobilized biomass reactors can be applied as an alternative technology for wastewater treatment that combines anaerobic and aerobic processes. Systems with immobilized biomass facilitate the use of more compact units operating without recirculation and separation systems. The effective control of cellular retention time, the possibility to achieve high biomass concentration and, consequently, the application of low hydraulic retention time are advantages that have stimulated the adoption of immobilized biomass technology [75].

A bench-scale radial anaerobic/aerobic immobilized biomass (RAAIB) reactor packed with polyurethane foam cubes was developed, with the aim of removing both organic matter and ammonium nitrogen from sanitary wastewater with average feed COD content of 345 mg/L. The reactor achieved an organic removal efficiency of 84% at HRT of 1.2–15.5 h [68].

As shown in Fig. 9, the RAAIB reactor is divided into five concentric chambers. The influent wastewater is fed at the top of the first chamber and flows radially from the anaerobic to the aerobic section. The second and fourth chambers are packed with 10-mm-sided polyurethane foam cubes. The second chamber is designed for anaerobic process and inoculated with anaerobic sludge while the fourth chamber is not. The third aerobic chamber contains eight porous stones which are distributed uniformly close to the bottom of the reactor and connected to a compressor to aerate and mix the wastewater. Finally the effluent is discharged from the bottom of the fifth chamber.

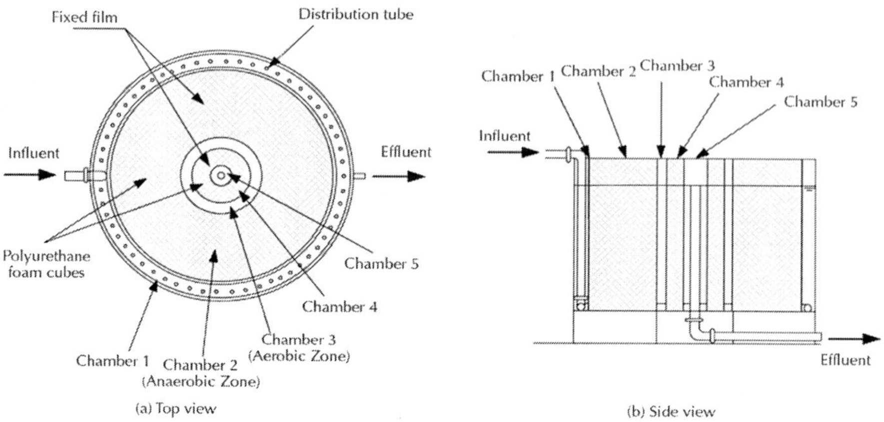

Figure 9: Schematic diagram of the RAAIB reactor a) Top view; b) Side view [68].

This configuration favors the transfer of oxygen to the liquid mass attributable to the fixed polyurethane foam bed arrangement in concentric chambers, and the reactor is easy to operate and control which make it suitable as an attractive option.

Simultaneous Aerobic–anaerobic (SAA) Bioreactor System

The simultaneous aerobic–anaerobic (SAA) bioreactor as shown in Fig. 10 is a combination of air lift reactor, fluidized bed and upflow anaerobic sludge blanket. There are an inner cylinder and an outer cylinder in the bioreactor, and the aerobic and anaerobic zones are established by controlling aeration location, aeration capacity and reactor shape. The aerobic zone is formed in the inner cylinder, as air is supplied from the bottom of the bioreactor. The anaerobic zone is formed in the outer cylinder due to limited oxygen transfer from the central region. The influent flows into the bottom of the bioreactor, and the resulting effluent is withdrawn from the top of the bioreactor. There is a decrease of dissolved oxygen concentration in the down flow zone as water flows from the inner zone to the outer zone; under oxygen-limited condition, aerobic and anaerobic process occurs simultaneously as a result of dissolved oxygen concentration gradients arising from diffusion limitations [76].

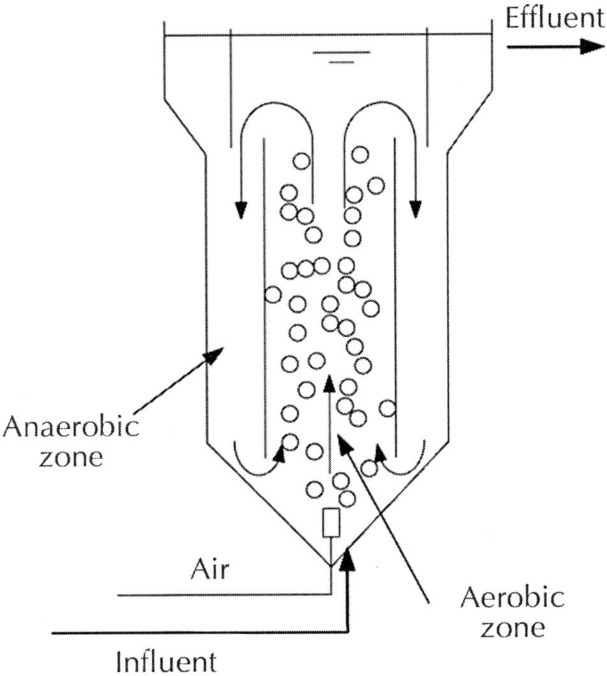

Figure 10: Schematic diagram of the SAA bioreactor system [70]. (Reprinted from Chemosphere, The biological treatment of landfill leachate using a simultaneous aerobic and anaerobic (SAA) bio-reactor system, 72 (2008) 1751–1756, with permission from the copyright holders, Elsevier).

The treatment of diluted landfill leachate wastewater with COD concentration varied from 1000 to 3300 mg/L was carried out in a SAA bioreactor system [70]. The reported COD removal efficiency ranged from 85 to 95%, with average removal efficiency of 94%.

The SAA bioreactor system is superior to conventional biological process for the effective removal of organic and nitrogenous matter from landfill leachates due to the reduced space requirement and operation management. Energy consumption is mainly utilized for the pump and air compressor and chemicals are rather rarely applied. Due to the simple operation and maintenance, the SAA bioreactor system is a good option for wastewater treatment, particularly in the developing countries.

Anaerobic–aerobic Integrative Baffled Bioreactor

An anaerobic–aerobic integrative baffled bioreactor was proposed for the treatment of potato starch processing wastewaters with COD values ranged from 1100 to 4500 mg/L [67]. In order to increase the efficiency, porous burnt-coke particles, a waste product of heavy industry is utilized as carriers in the aerobic zone to support the growth of microorganisms. With the presence of burnt-coke particles, the maximum COD reduction achieved was 98.7% while it was 96.0% in the absence of the burnt-coke. Burnt-coke carriers provide larger surface area for the attachment of the biofilm which leads to an increase in biomass concentration. The higher the biomass concentration in the carrier, the more organic matter was removed. An optimal HRT ranged from 12 to 24 h was used to produce effluent which was suitable for discharge.

The anaerobic–aerobic integrative baffled bioreactor is shown in Fig. 11 and includes three anaerobic zones, two depositions and one aerobic zone. It is rectangular and is subdivided equally into down flow and upflow sections by a series of 5-mm thick vertical high/low baffles. Due to the 45° turn out angle, the baffles cause the wastewater to rise and then flow downwards into the reactor. The system is divided into a three phase anaerobic biodegradation process. The first and second anaerobic zones are designed for hydrolysis, while the third anaerobic zone is mainly responsible for the production of methane. Depositions are designed for sedimentation and their main function is to separate the anaerobic and aerobic zone so that the anaerobic conditions are maintained in the anaerobic zones.

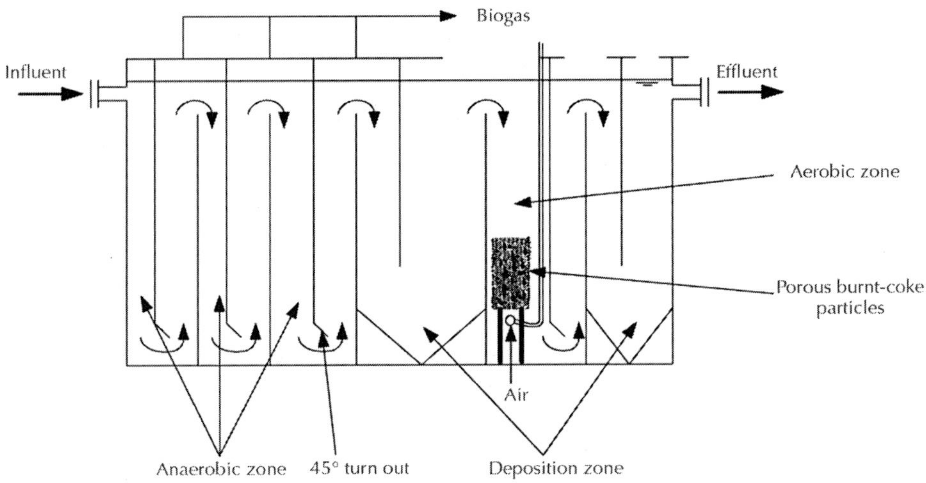

Figure 11: Schematic diagram of the anaerobic–aerobic integrative baffled bioreactor [67]. (Reprinted from Chemical Engineering Journal, Efficiency of porous burnt-coke carrier on treatment of potato starch wastewater with an anaerobic-aerobic bioreactor, 148 (2009) 35–40, with permission from the copyright holders, Elsevier).

The advantages of this bioreactor include rapid biodegradation, low yields of sludge and excellent process stability. This configuration is an effective solution to the treatment of wastewater for most small and medium-sized plants which possess little economic capacity to invest in environmental controls. Based on its influent COD, anaerobic COD removal percentage and theoretical conversion yield of methane gas (0.38 m^3CH_4/kg COD), it is estimated that a total of 1.23 tonnes CH_4/L influent wastewater can be produced in this bioreactor. It should be noted that 21 CERs, which is equivalent to 210 can be claimed and generated as revenue if 1 tonne of methane gas produced in the bioreactor are captured as renewable gas, based on carbon credit price of 10 per tonne of carbon [77]. Hence, in this case, approximately 260 per litre influent wastewater is highly potential of being awarded, which in turn will shorten the payback period on the investment of the bioreactor.

Integrated Bioreactors without Physical Separation of Anaerobic and Aerobic Zones

While the previous section covered anaerobic–aerobic treatment systems with separation between each process, a number of integrated bioreactors have been developed which allow the co-existence of anaerobic and aerobic populations inside the same reactor. This is done without physical separation using stacked configurations in which anaerobic conditions are maintained in the lower section while aerobic conditions are maintained in the upper part. This is achieved by introducing aeration at an intermediate height within the reactor [66], [78], [79], [80], [81], [82], [83] and [84] and Table 4 lists the systems based on this approach.

Upflow Anaerobic/Aerobic Fixed Bed (UA/AFB) Integrated Bioreactor

A bench scale upflow anaerobic/aerobic fixed bed (UA/AFB) integrated bioreactor as shown in Fig. 12 was developed and tested with synthetic wastewater. It is filled with PVC rings 1.5 cm in diameter as media and operated in an upflow mode that consists of two main zones: lower anaerobic zone and upper aerobic zone. Moosavi et al. [84] demonstrated that a total HRT of 9 h (5 h for anaerobic and 4 h for aerobic) is sufficient to accomplish efficient COD removal with more than 95% at OLR as high as 7.4 kg COD/m^3 d. The UA/AFB bioreactor is capable of handling high organic loads and able to recover immediately after any disturbances. The UA/AFB bioreactor is a potential biotechnology for treatment of industrial wastewater containing high organic loads. The study did not incorporate a methane gas capture system, which would produce biogas to partially offset the cost of treatment. It is projected that the bioreactor manages to attain approximately 20 carbon credit per liter of influent wastewater, which is equivalent to 200.

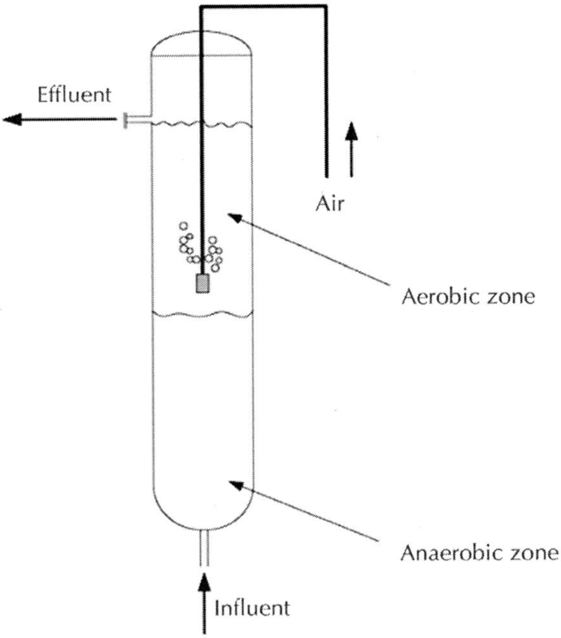

Figure 12: Schematic diagram of combined UA/AFB integrated reactor system [84].

Anaerobic–aerobic Granular Biofilm Bioreactor

Similar in design to a UA/AFB reactor, a granular biofilm bioreactor consists of an UASB with either an aeration column or a sparger placed in the middle part of the reactor. A schematic diagram of anaerobic–aerobic granular biofilm reactor is presented in Fig. 13. It has been utilized in the biodegradation of various chlorinated pollutants such as trichloethylene (TCE) [66] and [78] and polychlorinated biphenyl (PCB) [82]. The biodegradation of various chlorinated pollutants is based on the co-existence of aerobic methanotrophic and anaerobic methanogenic bacteria in a biofilm under oxygen-limited conditions.

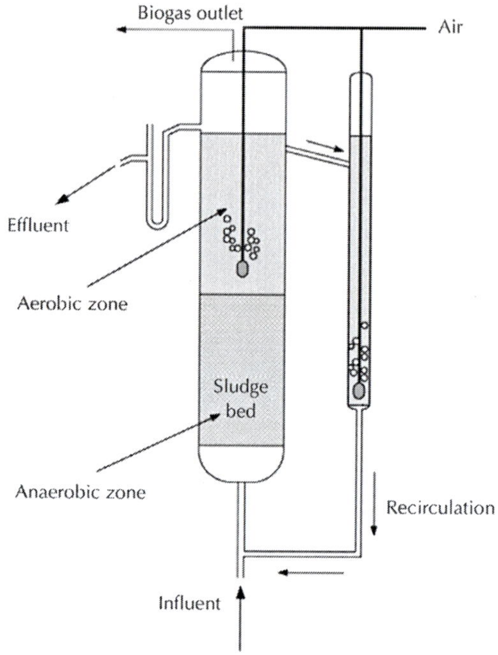

Figure 13: Schematic diagram of anaerobic–aerobic granular biofilm reactor [66] and [78]. (Reprinted from Biochemical Engineering Journal, Degradation of trichloroethylene in a coupled anaerobic-aerobic bioreactor: Modeling and experiment, 26 (2005) 72–81, with permission from the copyright holders, Elsevier).

Oxygen consumption by aerobic bacteria results in a steep oxygen gradient across the biofilm, leaving the interior a sufficiently thick biofilm free of oxygen and thereby provides a suitable niche for the growth of anaerobic methanogenic bacteria. Simultaneously, methane produced by the methanogens combined with the presence of oxygen favors the growth of aerobic methanotrophic bacteria in the outer layer of the biomass granules. Thus, anaerobic and aerobic populations of the biofilm co-exist closely in the same reactor system. It is a good strategy since both reductive and oxidative biotransformation occurs concomitantly to complete mineralization of highly substituted compounds under micro-aeration.

UASB reactors can accommodate low concentrations of oxygen without detrimental effects on the integrity or metabolic activity of the granular biomass. Thus, a partially aerated UASB reactor contains the substrates required by methanotrophic bacteria (i.e., indigenously produced methane and exogenously added oxygen) and could be an ideal system for maintaining consortia composed of methanogens and methanotrophs [78].

Shen and Guiot [79] investigated the impact of influent dissolved O_2, on the characteristics of anaerobic granular sludge at various dissolved O_2 concentrations (0.5–8.1 mg/L) via laboratory-scale anaerobic–aerobic granular biofilm bioreactor with a synthetic wastewater (carbon sources containing 75% sucrose and 25% acetate). As the granules are able to maintain good methanogenic activities when dissolved O_2 is present in the recirculated fluid, it indicates that the anaerobic–aerobic granular biofilm bioreactor can be successfully operated to maintain both active strict anaerobes and aerobes at the same time. With the elevated influent dissolved O_2, the methane yield declined from 64 to 42% of influent COD while the CO_2generation rate rose from 0.23 to 0.39 L (CO_2)/g COD, suggesting more organic substrate was aerobically mineralized under high dissolved O_2 conditions. However, in spite of significant aerobic COD elimination in the coupled reactors receiving high dissolved O_2 influent, a major part of the influent COD (at least 62%) was anaerobically removed.

However, the presence of dissolved O_2 in the recirculated fluid resulted in fluffy biolayers on the granule surface, which imposed a negative impact on the settleability of granular sludge and caused a slightly higher sludge washout. The negative impact of influent dissolved O_2 on the granule structure and settleability represent a drawback for the practical operation of the reactor.

Staged Anaerobic–aerobic Membrane Bioreactor (MBR)

A staged anaerobic–aerobic membrane bioreactor (MBR) in which the membrane module is submerged in the aerobic zone is shown in Fig. 14[83]. The aeration from the diffuser in the aerobic zone with the membrane module serves three purposes which are (i) providing oxygen for the biodegradation of substrates, (ii) mixing of the aerobic tank, and (iii) producing turbulence desirable to membrane cleaning. Porcelain carriers are installed to prevent the blockade of the orifice between the two zones of the reactor.

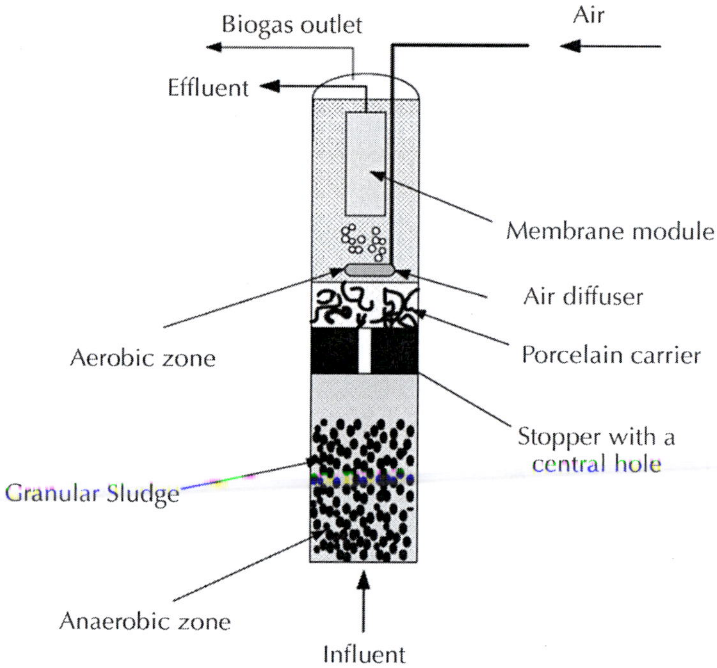

Figure 14: Schematic diagram of staged anaerobic–aerobic MBR [83]. (Reprinted from Process Biochemistry, The integration of methanogensis with simultaneous nitrification and denitrification in a membrane bioreactor, 40 (2005) 541–547, with permission from the copyright holders, Elsevier).

It has been employed successfully in the treatment of high strength synthetic wastewater containing high concentrations of ammonium with COD up to 10,500 mg/L and NH_4^+-N up to 1220 mg/L. The removal of the ammonium nitrogen from the high strength synthetic wastewater was accomplished through intermittent aeration in the aerobic zone, resulting in favorable conditions for the simultaneous nitrification and denitrification. The reported COD removals were exceeding 99% for OLR up to 10.08 kg COD/m^3 day. Between 60 and 80% of COD was anaerobically biodegraded in the anaerobic zone of the reactor and converted to methane that could serve as a carbon source for the denitrification in the aerobic zone. However, if the methane gas is captured as renewable energy, it is estimated that approximately 15 CER per liter of influent wastewater can be attained.

Integrated Anaerobic–aerobic Fixed-film Reactor (FFR)

An integrated pilot-scale anaerobic–aerobic fixed-film reactor (FFR) with arranged media, developed by Del Pozo and Diez [81], exhibited high performance on the removal of organic matter from slaughterhouse wastewater. Overall organic matter removal efficiencies of 93% were achieved for an average OLR of 0.77 kg COD/m^3 day at HRT of 0.94–3.8 days. It should be pointed out that the integrated anaerobic–aerobic FFR achieves higher treatment efficiency than the anaerobic–aerobic FFB system described in Section 3.4.

The anaerobic–aerobic FFR, shown in Fig. 15, consists of vertically oriented corrugated tubes while air is supplied using five independent membrane diffusers. The reactor is divided into two compartments, the aerobic zone (with aeration) and the anaerobic zone, without physical barriers. Wastewater enters the system from the upper part of the non-aerated region, through which it circulates downwards before being entrained up through the aerated zone due to the air-lift effect of the air injection. It then leaves the reactor from the upper part of the aerobic zone. Different anaerobic–aerobic volume ratios (V_{an}:V_{ae}) are achieved by turning on and off each diffuser at the bottom of the reactor.

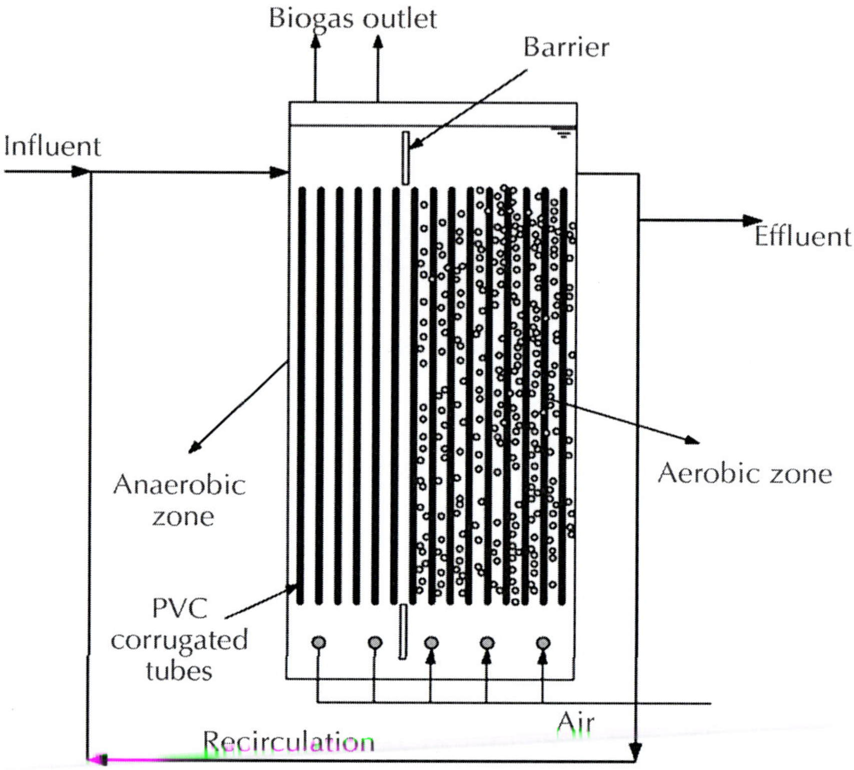

Figure 15: Schematic diagram of integrated anaerobic–aerobic FFB [85]. (Reprinted from Water Research, Integrated anaerobic-aerobic fixed-film reactor for slaughterhouse wastewater treatment, 39 (2005) 1114–1122, with permission from the copyright holders, Elsevier)

Most of the COD was removed through aerobic oxidation (96%), while the anoxic removal stood only for 2.6% and the methanogenic removal stood for 1.2% (V_{an}:V_{ae} was 3:2). When this ratio was reduced to 2:3 the COD removed by methanogenesis decreased to 0.6%. The main reason for the low extension of the anaerobic process is the high mixing pattern existing in the integrated reactor. High recirculation homogenizes the aerated and non-aerated zones, maintaining dissolved oxygen concentrations of 1.4 mg/l in the non-aerated zone. This significantly limits both the methanogenic and anoxic processes. Similar phenomenon was also reported by other authors working with bubble columns described in Section4.1.1. [69] And it is important to clearly separate the aerated and non-aerated zones that allow strict anaerobic conditions to appear. This is the reason for the inclusion of the two small barriers seen in Fig. 15at the top and at the bottom of the reactor. To recover the methane without dilution in the injected air, the aerated and non-aerated regions are set in parallel rather than in series.

Integrated Anaerobic–aerobic Fluidized Bed Reactor

Fdez-Polanco et al. [80] used a pilot scale anaerobic–aerobic fluidized bed reactor for the simultaneous elimination of organic carbon and nitrogen from municipal wastewater. Appreciable COD removal efficiencies (higher than 80%) were obtained at HRT of 24 h for an OLR of 1.2 kg COD/m^3 day.

Fig. 16 illustrates in schematic form the integrated anaerobic–aerobic fluidized bed reactor. The bioreactor is a cylindrical fluidized bed with pulverised pumice-stone as support material, and the aeration is carried out by four cylindrical fine bubble membrane diffusers arranged in a cross shape. This system is supported by a 'bell' (at the top of the diagram in Fig. 16) and different anaerobic/aerobic volume ratios (V_{an}:V_{ae}) can be achieved by varying its height in the interior of the bed.

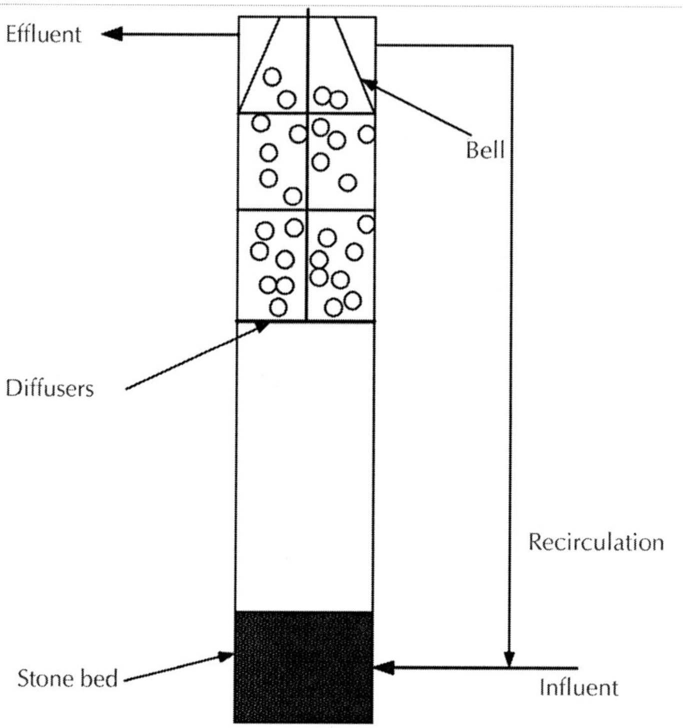

Figure 16: Schematic diagram of integrated anaerobic–aerobic fluidized reactor [80]. (Reprinted from Water Science & Technology, 29 (10–11), 339–346, with permission from the copyright holders, IWA).

The main advantages they observed were good stability despite variations in the organic load, a short start-up time, good recovery after unforeseen circumstances (e.g. the lack of aeration) and the possible automation using inexpensive technology. The main drawbacks identified were the requirement of additional pumping to maintain the support material in suspension. With the aim of diminishing the total COD and TSS of the effluent, the loss of support material as well as the concentration of dissolved oxygen in the effluent, a solid separation system in the effluent and recirculation should be implemented.

Comparison of Integrated Anaerobic–aerobic Bioreactor in Industrial Wastewater Treatment

While most of the integrated bioreactors discussed in this paper have not been implemented in an industry setting, the results on a laboratory, bench or pilot scale shown in Table 3 and Table 4 indicate they can be utilized to treat various types of industrial and municipal wastewater. They can deal with a wide range of COD concentration (ranging from 345 to 10,500 mg/L), and provide good quality effluent with at least 80% COD removal at HRT of 1.2 h to 3.8 days. However, biogas capture is not emphasized in those integrated bioreactors due to the uncertainties related to corrosive behavior of the biogas and lack of technical expertise. Additionally, some of the integrated bioreactors utilize biogas constituents as electron donor for denitrification instead of harnessing the generated biogas as renewable energy. The staged anaerobic–aerobic membrane bioreactor achieves the best COD removal (more than 99%) at high OLR and short HRT. However, while presenting the best performance it is considered a high cost from both a capital and operating perspective and suffers from problems associated with membrane fouling thereby frequent cleaning of the membrane is necessary.

The physical separation of anaerobic and aerobic zones provide a simplified design which aids construction, operation, control and maintenance when compared to equivalent bioreactors without physical separation. Separation also makes it easier to maintain strict anaerobic and aerobic conditions although it does lead to an increase in construction costs.

Using a stacked configuration in the design of integrated anaerobic–aerobic bioreactors is advantageous as it reduces space requirements, providing lower capital cost while maintaining excellent COD removal efficiencies of 95–98%. Furthermore, the biogas produced from the anaerobic zone can be captured as renewable energy from the intermediate height of the bioreactor with a stacked configuration, ensuring the methane produced is not oxidized in the aerobic zone. The use of a packing medium can also provide strict anaerobic and aerobic conditions to reduce the negative impacts of dissolved O_2 on the anaerobic granule sludge [79]. The addition of biomass to the small suspended medium in the integrated bioreactor allows sludge with high settling velocity and high biomass concentration which leads to a low-volume bioreactor. This permits compact integrated reactors and completely closed conditions.

Anaerobic–aerobic Sequencing Batch Reactor (SBR)

Conventional sequencing batch reactor (SBR) treatment systems generally consist of the batch steps of fill, react, settle, decant and idle in a cyclic operation with complete aeration during the react period to oxidize the organic matter and nitrify the ammonium-nitrogen of wastewater. Recently, SBR systems have been modified by adjusting the steps in the react cycle to provide anaerobic and aerobic phases in certain number and sequence for biological nutrient (C, N, and P) removal.

Most of the anaerobic–aerobic SBR systems are exploited in the treatment of textile wastewater for efficient color and COD removal [15], [85] and [86]. The enrichment of desired microbial population can be accomplished easily by alternating anaerobic–aerobic phases through the control of the aeration in the anaerobic–aerobic SBR system. The duration, oxygen concentration, and mixing can be altered according to the needs of the particular treatment plant. In the operation of an anaerobic–aerobic SBR, pure nitrogen gas is purged in the anaerobic phase while air is supplied in the aerobic phase.

Investigations have been done to observe the effects of anaerobic–aerobic residence time on the performance of SBR. The experimental results indicated that anaerobic and aerobic residences times in SBR system significantly affected the system's performance. In the textile wastewater treatment, a slightly higher COD removal of 90% with an anaerobic/aerobic cycle of 17.5/2.5 h was achieved compared to COD removal of 87% with an anaerobic/aerobic cycle of 14/6 h. It was found that the duration of the anaerobic phase should be long enough to obtain better COD and color removal [87].

Kapdan and Oztekin [86] showed similar results in the treatment of textile synthetic wastewater. The optimum anaerobic and aerobic residence times were determined as 12 and 11 h respectively with total reaction time of 23 h for efficient overall COD removal (more than 85%). When aerobic residence time (($_{\text{Haerobic}}$) is 19–20 h, COD removal efficiency was around 50% under anaerobic conditions and reached to about 80% by the contribution of the aerobic phase to COD removal. However, as the anaerobic residence time was increased, the contribution of aerobic phase on COD removal was negligible. This was the result of the toxic effect of the dyestuff biodegradation end products of the anaerobic phase. Due to the batch operation, these were accumulated in the system and long term exposure of the cultures to these products inhibited aerobic organism activity. While transition between anaerobic to aerobic phases restrains the growth of aerobic organisms, the facultative anaerobic culture needs longer aeration period to be active under aerobic conditions.

On the other hand, appreciable soluble COD and BOD_5 removal efficiencies of 85 ± 6% and 95 ± 4% were achieved respectively in the treatment of wool dyeing effluents. The optimum residence times are 8 h anaerobic reaction and 12 h aerobic reaction with total cycle of 24 h in fast filling mode. Results indicated a longer aeration time resulted in better performance, due to more efficient cell growth and COD combined with a fast fill that provided the feast-famine conditions that favored sludge settleability. However, a long aeration period is not feasible in the form of operational economy [85].

Anaerobic–aerobic SBR has proved to be a suitable technology for organic removal from textile wastewater as high COD removal of more than 85% is achieved. Hence, anaerobic–aerobic SBR shows great potential in the treatment of high strength industrial and municipal wastewater due to their simplicity in operation. However, further investigations on its control of anaerobic–aerobic microbial consortia, methanogenic activity, biomass yield and its ability to recover from shock organic loads are required.

Integrated Bioreactor Based on Combined Anaerobic–aerobic Cultures

Combined culture is the mixture of anaerobic and aerobic cultures that could survive under alternating anaerobic–aerobic conditions in the same reactor. Methanogenic and aerobic biological processes are often considered mutually exclusive and separated as biological wastewater treatment options. Dissolved oxygen (DO) even at low levels is considered to be extremely toxic to methanogens. Nonetheless, they have been found to survive short periods in the presence of dissolved oxygen and coexist with aerobic or microaerophilic organisms in a single mixed culture. The survival of anaerobic cultures under aerobic or microaerobic conditions (i.e. DO concentration < 1 mg/L) are due to the intrinsic tolerance or formation of anaerobic niches [88]. As a result, combined anaerobic–aerobic cultures have been investigated increasingly over the last two decades [88], [89], [90], [91], [92], [93], [94], [95] and [96].

Combined cultures have been applied successfully in the treatment of several contaminants such as polycyclic aromatic hydrocarbons, and highly chlorinated solvents that require sequentially operated anaerobic and aerobic or anoxic reactors [90], [94] and [95]. With free or co-immobilized cultures of anaerobes and aerobes, DO concentrations display alternating values. The oxygen gradient results in alternating conditions from aerobic to anaerobic either through the reactor content (as in packed bed or slurry reactors) or from bulk liquid to the depths of the immobilized cocultures. This leads to possible living conditions for different types of bacteria and makes the coexistence of anaerobic and aerobic cultures feasible. Interestingly, others have identified that with the addition of 4 mg

O_2/L day to essentially anaerobic cultures, methane production has been doubled when algae was the primary substrate [97].

For low strength municipal wastewaters treatment, combined cultures from a mixture of anaerobic granular and suspended aerobic cultures (40:60, v/v) were developed in an upflow sludge bed (USB) reactors. The combined cultures in USB reactor exhibited average BOD removal efficiency of 52–76% at HRT of 0.75 day. Combined cultures which were aerated every other day (i.e. alternating cyclic anaerobic to microaerobic/aerobic conditions) were considered as the optimum and feasible aeration protocol as compared to aeration for 4 h/day or continuously due to their higher removal efficiencies, slightly better settling characteristic and lower oxygen requirement [88].

Appreciable COD removal efficiencies (greater than 93%) were reported in the study of sucrose biotransformation under methanogenic and oxygen-limited conditions in bench-scale batch reactors seeded with a mixture containing anaerobic digester sludge and aerobic mixed liquor. In addition to oxygen-limited reactors, anaerobic (methanogenic) and aerobic (dissolved oxygen greater than 2.0 mg/l) systems were operated in parallel for comparison. It was observed that the overall COD removal efficiencies for oxygen-limited cultures, strictly anaerobic cultures, and strictly aerobic cultures were comparable under the complete-mix, suspended growth conditions. The limited-aeration conditions were achieved by introducing air through a timer-actuated solenoid valve which was open for 15 seconds in every half-hour [91].

Integrated bioreactors containing anaerobic–aerobic cultures can be beneficial compared to conventional anaerobic and aerobic treatment system due to their achievement of required discharge standards, prevention of biomass loss, high settling characteristic, reduced aeration and additional methane production. Combined cultures are perceived as an energy efficient treatment alternative to achieve low final COD concentrations, minimal biosolids generation, and mineralization of a broad range of specific organic chemicals. Nonetheless, further research is required to address the application and process control issues of these types of bioreactor systems.

Overall Comparison of the Integrated Anaerobic–aerobic Bioreactors

The advantages and disadvantages of the four types of integrated anaerobic–aerobic bioreactors are presented in Table 5. The anaerobic–aerobic SBR and combined anaerobic–aerobic culture system require less energy input than either types of integrated bioreactors, as less aeration is required. However, inappropriate control of anaerobic and aerobic residences times in SBR system will result in difficulties controlling the anaerobic–aerobic microbial consortia and thus the selection and enrichment of the desired microbial population will be difficult to achieve. A high level of sophistication is required in timing units, controller, and software and censors while automated switches and valves are required to achieve efficient organic removal in anaerobic–aerobic SBR system.

Table 5: Advantages and disadvantages of various types of integrated anaerobic–aerobic bioreactors

Bioreactor Type	Advantages	Disadvantages
Integrated bioreactors with and without physical separation of anaerobic–aerobic zone	High organic removal efficiency	Complicated design of bioreactor
	Capable of handling high OLR	Relatively higher construction cost
Anaerobic–aerobic SBR	Single tank configuration without the need of clarifier	Complex control of anaerobic–aerobic microbial consortia
	Low capital cost	Require special attention to determine anaerobic and aerobic residence time
	Low energy requirement	High level of sophistication is required
	Flexibility in operation	
Combined anaerobic–aerobic culture system	Low energy requirement	Complex control of anaerobic–aerobic microbial consortia
	Small reactor volume	Sensitive to environmental condition

For combined anaerobic–aerobic culture system, control of aeration time and maintenance of oxygen limited condition in the bioreactor are critical as methanogens are sensitive to the presence of dissolved oxygen. Excessive dissolved oxygen concentration will restrain methanogenic activity and adversely affect the performance of the system.

Integrated bioreactors with and without physical separation of anaerobic–aerobic zone have exhibited the best performance in terms of COD removal and are capable of handling high OLR when compared to the anaerobic–aerobic SBR and combined anaerobic–aerobic culture system. However, the design and configuration of integrated bioreactor are more complicated and therefore, its construction cost inevitably would be higher than anaerobic–aerobic SBR system and combined anaerobic–aerobic culture system. With proper installation of methane gas capturing system, integrated anaerobic–aerobic bioreactor emerges as a viable technology in treating high strength industrial wastewaters, in which it offers an attractive energy recovery source while reducing greenhouse gas emissions. This may in the future lead to it being awarded carbon credits from the Clean Development Mechanism (CDM), further shortening the payback period for investment.

CONCLUSIONS

Anaerobic–aerobic treatments receive great attention over the past decades due to their numerous advantages such as low energy consumption, low chemical consumption, low sludge production, vast potential of resource recovery, less equipment required and high operational simplicity. However, conventional anaerobic–aerobic systems are found to have operational limitations in terms of long HRT, space requirement and facilities to capture biogas. The applications of newly developed high rate bioreactors address these limitations and provide increased organic matter removal at shorter HRT and higher methane yields for biogas production.

In order to meet strict constraints with respect to space, odors and minimal sludge production, considerable attention has been directed towards the integrated anaerobic–aerobic bioreactors which combine the aerobic and anaerobic process in a single bioreactor. With simple yet cost effective technology, the generation of renewable energy and outstanding treatment efficiency, it is envisaged that the compact integrated bioreactors will be able to treat a wide range of high organic strength industrial and municipal wastewater. However, most of the integrated bioreactors reported in this work lack large scale implementation within industry and further work is required to evaluate the performance of these promising reactors on a larger scale. Besides, further improvements such as installation of biogas capture system and utilization of suspended carrier or packing medium are considered essential.

As such, the integrated bioreactors can attain carbon credit derived from Clean Development Mechanism (CDM) under Kyoto Protocol 1997, shifting the paradigm of wastewater management from 'treatment and disposal' to 'beneficial utilization' as well as 'profitable endeavor'.

REFERENCES

1. W.J. Ng, Industrial Wastewater Treatment, World Scientific Publishing Company, 2006.
2. Metcalf and Eddy, Wastewater Engineering Treatment and Reuse, fourth ed., McGraw Hill, 2003.
3. D.T. Sponza, A. Uluköy, Treatment of 2,4-dichlorophenol (DCP) in a sequential anaerobic (upflow anaerobic sludge blanket) aerobic (completely stirred tank) reactor system, Process Biochemistry 40 (2005) 3419–3428.
4. B.G. Yeoh, Anaerobic treatment of industrial wastewaters in Malaysia, in: Post Conference Seminar on Industrial Wastewater Management in Malaysia, Kuala Lumpur, Malaysia, 1995.
5. L. Seghezzo, G. Zeeman, J.B. van Lier, H.V.M. Hamelers, G. Lettinga, A review: The anaerobic treatment of sewage in

UASB and EGSB reactors, Bioresource Technology 65 (1998) 175–190.

6. F.Y. Cakir, M.K. Stenstrom, Greenhouse gas production: a comparison between aerobic and anaerobic wastewater treatment technology, Water Research 39 (2005) 4197–4203.
7. C.P. Leslie Grady Jr., G.T. Daigger, H.C. Lim, Biological Wastewater Treatment, second ed., revised and expanded, CRC Press, 1999.
8. J.J. Heijnen, A. Mulder, R. Weltevrede, J. Hols, H. Vanleeuwen, Large-scale anaerobic–aerobic treatment of complex industrial-waste water using biofilm reactors, Water Science and Technology 23 (1991) 1427–1436.
9. N.F. Gray, Water Technology: An Introduction for Environmental Scientists and Engineers, Elsevier, Oxford, 2005.
10. G.G. Aggelis, H.N. Gavala, G. Lyberatos, Combined, separate aerobic and anaerobic biotreatment of green olive debittering wastewater, Journal of Agricultural Engineering Research 80 (2001) 283–292.
11. M. Vera, E. Aspe, M.C. Marti, M. Roeckel, Optimization of a sequential anaerobic–aerobic treatment of a saline fishing effluent, Process Safety and Environmental Protection 77 (1999) 275–290.
12. B. Frostell, Anaerobic–Aerobic Biological Treatment of Starch Industry Waste Waters, Starch—Stärke 35 (1983) 185–189.
13. F.J. Cervantes, S.G. Pavlostathis, A.C. van Haandel, Advanced Biological Treatment Processes for Industrial Wastewaters: Principles and Applications, IWA Publishing, 2006.
14. M. Ros, G.D. Zupancic, Two-stage thermophilic anaerobic–aerobic digestion of waste-activated sludge, Environmental Engineering Science 21 (2004) 617–626.
15. N. Supaka, K. Juntongjin, S. Damronglerd, M.-L. Delia, P. Strehaiano, Microbial decolorization of reactive azo dyes in a sequential anaerobic–aerobic system, Chemical Engineering Journal 99 (2004) 169–176.

16. H. Liu, C. Yang, W. Pu, J. Zhang, Removal of nitrogen from wastewater for reusing to boiler feed-water by an anaerobic/aerobic/membrane bioreactor, Chemical Engineering Journal 140 (2008) 122–129.
17. L.K. Wang, Waste Treatment in the Process Industries, CRC Press, 2005.
18. Y.T. Ahn, S.T. Kang, S.R. Chae, C.Y. Lee, B.U. Bae, H.S. Shin, Simultaneous high-strength organic and nitrogen removal with combined anaerobic upflow bed filter and aerobic membrane bioreactor, Desalination 202 (2007) 114– 121.
19. Y. Zhang, L. Yan, L. Chi, X. Long, Z. Mei, Z. Zhang, Startup and operation of anaerobic EGSB reactor treating palm oil mill effluent, Journal of Environmental Sciences 20 (2008) 658–663.
20. O.N. Agdag, D.T. Sponza, Anaerobic/aerobic treatment of municipal landfill leachate in sequential two-stage up-flow anaerobic sludge blanket reactor (UASB)/completely stirred tank reactor (CSTR) systems, Process Biochemistry 40 (2005) 895–902.
21. K.V. Lo, P.H. Liao, Anaerobic–aerobic biological treatment of a mixture of cheese whey and dairy manure, Biological Wastes 28 (1989) 91–101.
22. Y. Zhang, L. Yan, X. Qiao, L. Chi, X. Niu, Z. Mei, Z. Zhang, Integration of biological method and membrane technology in treating palm oil mill effluent, Journal of Environmental Sciences 20 (2008) 558–564.
23. E.J. La Motta, E. Silva, A. Bustillos, H. Padron, J. Luque, Combined anaerobic/aerobic secondary municipal wastewater treatment: Pilot-plant demonstration of the UASB/aerobic solids contact system, Journal of Environmental Engineering—ASCE 133 (2007) 397–403.
24. A. Castillo, P. Llabres, J. Mata-Alvarez, A kinetic study of a combined anaerobic–aerobic system for treatment of domestic sewage, Water Research 33 (1999) 1742–1747.

25. J.-h. Im, H.-j. Woo, M.-w. Choi, K.-b. Han, C.-w. Kim, Simultaneous organic and nitrogen removal from municipal landfill leachate using an anaerobic–aerobic system, Water Research 35 (2001) 2403–2410.

26. M. Isik, D.T. Sponza, Biological treatment of acid dyeing wastewater using a sequential anaerobic/aerobic reactor system, Enzyme and Microbial Technology 38 (2006) 887–892.

27. M. Isik, D.T. Sponza, Anaerobic/aerobic sequential treatment of a cotton textile mill wastewater, Journal of Chemical Technology and Biotechnology 79 (2004) 1268–1274.

28. M. Isik, D.T. Sponza, Anaerobic/aerobic treatment of a simulated textile wastewater, Separation and Purification Technology 60 (2008) 64–72.

29. U. Tezel, E. Guven, T.H. Erguder, G.N. Demirer, Sequential (anaerobic/aerobic) biological treatment of dalaman SEKA Pulp and Paper Industry effluent, Waste Management 21 (2001) 717–724.

30. D.T. Sponza, P. Demirden, Treatability of sulfamerazine in sequential upflow anaerobic sludge blanket reactor (UASB)/ completely stirred tank reactor (CSTR) processes, Separation and Purification Technology 56 (2007) 108–117.

31. N. Gizgis, M. Georgiou, E. Diamadopoulos, Sequential anaerobic/aerobic biological treatment of olive mill wastewater and municipal wastewater, Journal of Chemical Technology and Biotechnology 81 (2006) 1563–1569.

32. V. Sklyar, A. Epov, M. Gladchenko, D. Danilovich, S. Kalyuzhnyi, Combined biologic (anaerobic–aerobic) and chemical treatment of starch industry wastewater, Applied Biochemistry and Biotechnology 109 (2003) 253– 262.

33. M. von Sperling, V.H. Freire, C.A. de Lemos Chernicharo, Performance evaluation of a UASB-activated sludge system treating municipal wastewater, Water Science and Technology 43 (2001) 323–328.

34. H. Chen, ATP content and biomass activity in sequential anaerobic/aerobic reactors, Journal of Zhejiang University Science 5 (2004) 727–732.
35. J. Yu, H. Chen, M. Ji, P.L. Yue, Distribution and change of microbial activity in combined UASB and AFB reactors for wastewater treatment, Bioprocess Engineering 22 (2000) 315–322.
36. K.V. Lo, P.H. Liao, Anaerobic aerobic biological treatment of screened dairy manure, Biomass 10 (1986) 187–193.
37. R. Del Pozo, V. Diez, Organic matter removal in combined anaerobic–aerobic fixed-film bioreactors, Water Research 37 (2003) 3561–3568.
38. G.R. Pophali, R. Khan, R.S. Dhodapkar, T. Nandy, S. Devotta, Anaerobic–aerobic treatment of purified terephthalic acid (PTA) effluent; a techno-economic alternative to two-stage aerobic process, Journal of Environmental Management 85 (2007) 1024–1033.
39. R.H. Kettunen, J.A. Rintala, Sequential anaerobic–aerobic treatment of sulfurrich phenolic leachates, Journal of Chemical Technology and Biotechnology 62 (1995) 177–184.
40. I.K. Kapdan, S. Alparslan, Application of anaerobic–aerobic sequential treatment system to real textile wastewater for color and COD removal, Enzyme and Microbial Technology 36 (2005) 273–279.
41. M. von Sperling, C.A. de Lemos Chernicharo, Biological Wastewater Treatment in Warm Climate Regions, IWA Publishing, 2005.
42. A. Vlyssides, E.M. Barampouti, S. Mai, Influence of ferrous iron on the granularity of a UASB reactor, Chemical Engineering Journal 146 (2009) 49–56.
43. A. Torkian, A. Eqbali, S.J. Hashemian, The effect of organic loading rate on the performance of UASB reactor treating slaughterhouse effluent, Resources Conservation and Recycling 40 (2003) 1–11.

44. G.-P. Sheng, H.-Q. Yu, H. Cui, Model-evaluation of the erosion behavior of activated sludge under shear conditions using a chemical-equilibrium-based model, Chemical Engineering Journal 140 (2008) 241–246.
45. M. Lerner, N. Stahl, N. Galil, Aerobic vs. anaerobic–aerobic biotreatment: Paper mill wastewater, Environmental Engineering Science 24 (2007) 277– 285.
46. W.K. Shieh, J.D. Keenan, Fluidized bed biofilm reactor for wastewater treatment, Advances in Biochemical Engineering Biotechnology 33 (1986) 131–169.
47. A.A. Ulson de Souza, H.L. Brandão, I.M. Zamporlini, H.M. Soares, S.M.D.A. Guelli Ulson de Souza, Application of a fluidized bed bioreactor for cod reduction in textile industry effluents, Resources, Conservation and Recycling 52 (2008) 511–521.
48. J.J. Heijnen, A. Mulder, W. Enger, F. Hoeks, Review on the application of anaerobic fluidized bed reactors in waste-water treatment, The Chemical Engineering Journal 41 (1989) B37–B50.
49. V. Lazarova, J. Manem, Advances in biofilm aerobic reactors ensuring effective biofilm activity control, Water Science and Technology 29 (1994) 319– 327.
50. R. Saravanane, D.V.S. Murthy, Application of anaerobic fluidized bed reactors in wastewater treatment: a review, Environmental Management and Health 11 (2000) 97–117.
51. C.R.G. Tavares, G.L. Santanna, B. Capdeville, The effect of air superficial velocity on biofilm accumulation in a 3-phase fluidized-bed reactor, Water Research 29 (1995) 2293–2298.
52. A.C. Yeh, C. Lu, M. Lin, Performance of an anaerobic rotating biological contactor: effects of flow rate and influent organic strength, Water Research 31 (1997) 1251–1260.
53. Y.-H. Kim, C. Yoo, I.-B. Lee, Optimization of biological nutrient removal in a SBR using simulation-based iterative dynamic programming, Chemical Engineering Journal 139 (2008) 11–19.

54. S.V. Mohan, N.C. Rao, K.K. Prasad, B.T.V. Madhavi, P.N. Sharma, Treatment of complex chemical wastewater in a sequencing batch reactor (SBR) with an aerobic suspended growth configuration, Process Biochemistry 40 (2005) 1501–1508.
55. Y. Gao, C.P. Feng, C. Liu, Municipal wastewater treatment using sequencing batch biofilm reactor (SBBR), in: International Conference on Advances in Chemical Technologies for Water and Wastewater Treatment, Xian, PR China, 2008.
56. M. Kabacinski, B. Hultman, E. Plaza, J. Trela, Strategies for improvement of sludge quality and process performance of sequencing batch reactor plant treating municipal and tannery wastewater, Water Science and Technology 38 (1998) 69–77.
57. B. Ni, W. Xie, S. Liu, H. Yu, Y. Wang, G. Wang, X. Dai, Granulation of activated sludge in a pilot-scale sequencing batch reactor for the treatment of low-strength municipal wastewater, Water Research 43 (2009) 751–761.
58. P.L. Bishop, Biofilm structure and kinetics, Water Science and Technology 36 (1997) 287–294.
59. A.V. Haandel, J.V.D. Lubbe, Handbook Biological Waste Water Treatment: Design and Optimisation of Activated Sludge Systems, Anaerobic–aerobic Wastewater Treatment, 2007.
60. C. Wen, X. Huang, Y. Qian, Domestic wastewater treatment using an anaerobic bioreactor coupled with membrane filtration, Process Biochemistry 35 (1999) 335–340.
61. T. Ueda, N.J. Horan, Fate of indigenous bacteriophage in a membrane bioreactor, Water Research 34 (2000) 2151–2159.
62. E.B. Muller, A.H. Stouthamer, H.W. van Verseveld, D.H. Eikelboom, Aerobic domestic waste water treatment in a pilot plant with complete sludge retention by cross-flow filtration, Water Research 29 (1995) 1179–1189.
63. H. Dhaouadi, B. Marrot, Olive mill wastewater treatment in a membrane bioreactor: Process feasibility and performances, Chemical Engineering Journal 145 (2008) 225–231.

64. Z. Wang, Z. Wu, Distribution and transformation of molecular weight of organic matters in membrane bioreactor and conventional activated sludge process, Chemical Engineering Journal 150 (2009) 396–402.
65. N. Cicek, H. Winnen, M.T. Suidan, B.E. Wrenn, V. Urbain, J. Manem, Effectiveness of the membrane bioreactor in the biodegradation of high molecular weight compounds, Water Research 32 (1998) 1553–1563.
66. B. Tartakovsky, M.F. Manuel, S.R. Guiot, Degradation of trichloroethylene in a coupled anaerobic–aerobic bioreactor: Modeling and experiment, Biochemical Engineering Journal 26 (2005) 72–81.
67. R.-M. Wang, Y. Wang, G.-P. Ma, Y.-F. He, Y.-Q. Zhao, Efficiency of porous burnt-coke carrier on treatment of potato starch wastewater with an anaerobic–aerobic bioreactor, Chemical Engineering Journal 148 (2009) 35–40.
68. L.H.P. Garbossa, K.R. Lapa, M. Zaiat, E. Foresti, Development and evaluation of a radial anaerobic/aerobic reactor treating organic matter and nitrogen in sewage, Brazilian Journal of Chemical Engineering 22 (2005) 511–519.
69. T. Hano, M. Matsumoto, K. Kuribayashi, Y. Hatate, Biological nitrogen removal in a bubble column with a draught tube, Chemical Engineering Science 47 (1992) 3737–3744.
70. Z. Yang, S. Zhou, The biological treatment of landfill leachate using a simultaneous aerobic and anaerobic (SAA) bio-reactor system, Chemosphere 72 (2008) 1751–1756.
71. Y. Bando, T. Kato, K. Yasuda, Y. Sakurai, M. Nakamura, Wastewater treatment by anaerobic–aerobic activated sludge method in bubble column with draft tube, Journal of Chemical Engineering of Japan 32 (1999) 770–775.
72. L.G. Meng, Y. Bando, M. Nakamura, Development of rectangular airlift bubble column installed with support material for enhancement of nitrogen removal, Journal of Bioscience and Bioengineering 98 (2004) 269–273.

73. L.G. Meng, N. Tsuno, Y. Bando, M. Nakamura, Enhancement of nitrogen removal in a rectangular airlift bubble column having aerobic and anaerobic regions by adding an immobilizing carrier, Journal of Chemical Engineering of Japan 37 (2004) 399–405.

74. L.G. Meng, Y. Bando, M. Nakamura, Dissolved oxygen distribution in a rectangular airlift bubble column for biological nitrogen removal, Journal of Chemical Engineering of Japan 37 (2004) 1005–1011.

75. A. Fazolo, E. Foresti, M. Zaiat, Removal of nitrogen and organic matter in a radial- flow aerobic-anoxic immobilized Biomass reactor used in the post treatment of anaerobically treated effluent, Applied Biochemistry and Biotechnology 142 (2007) 44–51.

76. N. Puznava, M. Payraudeau, D. Thornberg, Simultaneous nitrification and denitrification in biofilters with real time aeration control, Water Science and Technology 43 (2001) 269–276.

77. R.N. Menon, Dialogue session with the palm oil industry and stakeholders, Palm oil Engineering Bulletin 83 (2007) 11–14.

78. C.B. Miguez, C.F. Shen, D. Bourque, S.R. Guiot, D. Grouleau, Monitoring methanotrophic bacteria in hybrid anaerobic–aerobic reactors with PCR and a catabolic gene probe, Applied and Environmental Microbiology 65 (1999) 381– 388.

79. C.F. Shen, S.R. Guiot, Long-term impact of dissolved O2 on the activity of anaerobic granules, Biotechnology and Bioengineering 49 (1996) 611–620.

80. F. Fdez-Polanco, F.J. Real, P.A. Garcia, Behavior of an anaerobic/aerobic pilotscale fluidized-bed for the simultaneous removal of carbon and nitrogen, Water Science and Technology 29 (1994) 339–346.

81. R. Del Pozo, V. Diez, Integrated anaerobic–aerobic fixed-film reactor for slaughterhouse wastewater treatment, Water Research 39 (2005) 1114–1122.

82. B. Tartakovsky, A. Michotte, J.C.A. Cadieux, P.C.K. Lau, J. Hawari, S.R. Guiot, Degradation of Aroclor 1242 in a single-stage coupled anaerobic/aerobic bioreactor, Water Research 35 (2001) 4323–4330.
83. D. Zhang, P. Lu, T. Long, W. Verstraete, The integration of methanogensis with simultaneous nitrification and denitrification in a membrane bioreactor, Process Biochemistry 40 (2005) 541–547.
84. G.R. Moosavi, A.R. Mesdaghinia, K. Naddafi, A.H. Mahvi, Feasibility of development and application of an up-flow anaerobic/aerobic fixed bed combined reactor to treat high strength wastewaters, in: 10th World Congress on Anaerobic Digestion, Montreal, Canada, 2004, pp. 2436–2440.
85. I.C. Goncalves, S. Penha, M. Matos, A.R. Santos, F. Franco, H.M. Pinheiro, Evaluation of an integrated anaerobic/aerobic SBR system for the treatment of wool dyeing effluents, Biodegradation 16 (2005) 81–89.
86. I.K. Kapdan, R. Oztekin, The effect of hydraulic residence time and initial COD concentration on color and COD removal performance of the anaerobic–aerobic SBR system, Journal of Hazardous Materials 136 (2006) 896–901.
87. N. Pasukphun, S. Vinitnantharat, Degradation of organic substances and reactive dye in an immobilized-cell sequencing batch reactor operation on simulated textile wastewater, in: 5th International Conference on Small Water and Wastewater Treatment Systems, Istanbul, Turkey, 2002.
88. T.H. Erguder, G.N. Demirer, Low-strength wastewater treatment with combined granular anaerobic and suspended aerobic cultures in upflow sludge blanket reactors, Journal of Environmental Engineering—ASCE 134 (2008) 295–303.
89. H. Kurosawa, H. Tanaka, Advances in immobilized cell culture—development of co-immobilized mixed culture system of aerobic and anaerobic microorganisms, Process Biochemistry 25 (1990) 189–196.
90. J. Gerritse, J.C. Gottschal, Mineralization of the herbicide 2,3,6-trichlorobenzoic acid by a co-culture of anaerobic and

aerobic-bacteria, FEMS Microbiology Ecology 101 (1992) 89–98.

91. D.H. Zitomer, J.D. Shrout, Feasibility and benefits of methanogenesis under oxygen-limited conditions, Waste Management 18 (1998) 107–116.
92. L. Yerushalmi, M.F. Manuel, S.R. Guiot, Biodegradation of gasoline and BTEX in a microaerophilic biobarrier, Biodegradation 10 (1999) 341–352.
93. H. Gardin, J.M. Lebeault, A. Pauss, Degradation of 2,4,6-trichlorophenol (2,4,6- TCP) by co-immobilization of anaerobic and aerobic microbial communities in an upflow reactor under air-limited conditions, Applied Microbiology and Biotechnology 56 (2001) 524–530.
94. J. Beunink, H.J. Rehm, Coupled reductive and oxidative-degradation of 4- chloro-2-nitrophenol by a co-immobilized mixed culture system, Applied Microbiology and Biotechnology 34 (1990) 108–115.
95. J. Beunink, H.J. Rehm, Synchronous anaerobic and aerobic degradation of DDT by an immobilized mixed culture system, Applied Microbiology and Biotechnology 29 (1988) 72–80.
96. J. Gerritse, F. Schut, J.C. Gottschal, Mixed chemostat cultures of obligately aerobic and fermentative or methanogenic bacteria grown under oxygen-limiting conditions, FEMS Microbiology Letters 66 (1990) 87–89.
97. S.J. Pirt, Y.K. Lee, Enhancement of methanogenesis by traces of oxygen in bacterial digestion of biomass, FEMS Microbiology Letters 18 (1983) 61–63.

Chapter 2

Combining "Chimie douce" and Green Principles for the Developing World: Improving Industrial Viability of Photocatalytic Water Remediation

Tiju Thomas[a] and Nagaraju Kottam[b]

[a]Materials Research Centre, Indian Institute of Science, Bangalore 560012, Karnataka, India

[b]Department of Chemistry, M. S. Ramaiah Institute of Technology, Bangalore 560054, Karnataka, India

ABSTRACT

We highlight the need for a comprehensive, multi-disciplinary approach for the development of cost-effective water remediation methods. Combining "chimie douce" and green chemical principles seems essential for making these technologies economically viable and socially relevant (especially in the developing world). A comprehensive approach to water remediation will take into account issues such as nanotoxicity, chemical yield, cost, and ease of deployment in reactors. By considering technological challenges that lie ahead, we will attempt to identify directions that are likely to make photocatalytic water remediation a more global technology than it currently is.

INTRODUCTION

Water treatment is of paramount importance in an increasingly populous world. Currently a whopping 80% of diseases in the developing world are linked to lack of availability of clean water (Centre, 2009a). If the current trend continues, 1.8 billion people will be living in regions with water scarcity (<1700 m^3 of water per person, per year) (Centre, 2009b). Averting such a situation through suitable water treatment technologies presents an urgent engineering challenge. Viable solutions will require insights from materials chemistry, semiconductor photochemistry, green chemistry, environmental science, toxicology and process engineering (Bahnemann, 2000).

Photocatalytic water remediation has long been seen as an option for detoxifying and sanitizing water using sunlight. The primary appeal of this technology is that sunlight is both green and free. Recent developments in visible light photocatalysis (Yu et al., 2012 and Zhang et al., 2011) makes this technology particularly appealing since visible light accounts for ~97% of energy in solar radiation. Currently the number of academic reports on novel photocatalysts continues to grow at an impressive pace (Khin et al., 2012). In any case, considering the long history of photocatalysis, the impact of academic research on industry has not been as impressive as it was once thought to be. Perhaps it is time to consider directions that will allow translation of lab technologies in this area. In particular, chemical engineers and industrial chemists may note that this technology has had negligible (if any) impact in the developing world.

Many countries in the developing world are solar rich and in principle, should be "hot spots" for solar based technologies such as photocatalytic water remediation and photovoltaics. The reality however is a far cry from this imagined scenario. What could have gone wrong? How is it that these novel and enabling water-treatment technologies have failed to impact the developing world? What is it that chemists and engineers can do to address this problem? We will try to answer these questions, and seek directions may make much of our research socially relevant. We will pay attention to trends in photocatalysis literature, and try to determine "loopholes" in our current research philosophy, which might be holding us back.

THE PROMISE OF PHOTOCATALYTIC WATER REMEDIATION: EXPECTATIONS AND THE REALITY

A particular promise of photocatalyst is its ability to sanitize (disinfect) water (Fig. 1). Sanitary quality of water is often assessed using bacteriological studies, wherein the concentration of Coliform in water is assayed. There is no easy access to toilets in many parts of the developing world (Nations, 2008), which make the fresh water sources particularly vulnerable to microbial infection, and hence high in Coliform concentration. In fact nearly all deaths (~99%) in the developing world occur due to water, sanitation and hygiene related causes (Prüss-Üstün et al., 2008). The extent of damage this lack of access to clean water creates is hard to fathom. An example may be helpful: roughly 20% of child deaths are due to diarrhea, which amounts to a loss of about 1.5 million lives each year (UNESCO/WHO, 2009). Sanitary quality of water is a global issue, and requires a technological solution that is clean, affordable, and easily implementable.

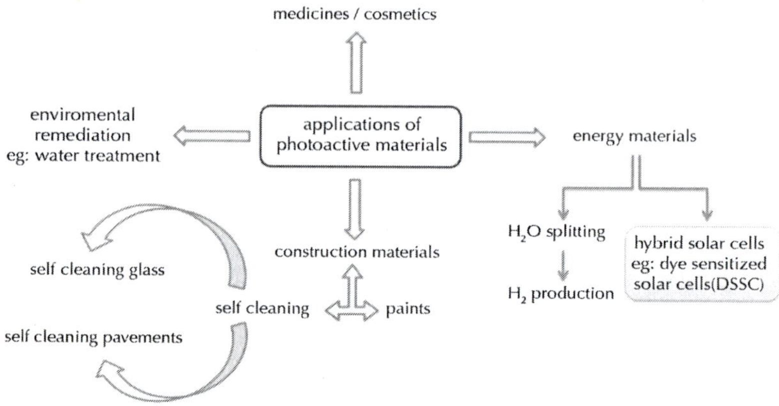

Figure 1: Photoactive materials are useful for a wide range of applications. The fields where they are applicable include medicine and cosmetics, energy materials, environmental remediation, and construction materials.

Apart from disinfection of water, photocatalysis holds promise in several other areas of social relevance (Fig. 1). For example, it is also a promising method for detoxification and decolorization of water. If synthesis of photocatalysts and implementation of photoreactors becomes cost-effective, then the technology has much to offer. However the opacity and turbidity of waste water, combined with its sheer volume in most modern cities pose a unique challenge to this technology.

There is reason to believe that viable water remediation solutions in modern cities and townships will require a multi-pronged approach (Bahnemann, 2000). Such an approach will most likely require use of a multi-stage reactor wherein one of the stages is dedicated to photocatalytic remediation. The process design and allied engineering aspects of such a photoreactor is still a matter of considerable research (McCullagh et al., 2011). In the long run photocatalysis is expected to minimize the energy requirement for comprehensive water treatment. It is hard to see how photocatalysis alone will supplant every other mature water treatment technology currently in place. Given that much of the waste water is opaque, a photocatalytic reactor can only be used after a few stages of coagulation, filtration etc. Table 1 gives a quick, but useful survey of current water treatment technologies (Chan et al., 2006, Chong et al., 2010 and Pérez-González et al., 2012) including those that will be employed prior to photocatalytic purification (Fig. 2). Table 1 is expected to provide a quick survey of advantages, challenges, and disadvantages of purification strategies currently in use.

Table 1: Current water treatment technologies: method, operating principle(s), performance metrics, advantages, disadvantages, and general remarks (Chan et al., 2006, Chong et al., 2010 and Pérez-González et al., 2012)

Physical/chemical method	**Operating principles/ conditions**	**Performance metric**	**Advantages**	**Disadvantages**	**Remarks**
Coagulation and precipitation	Coagulation and flocculation arising out of addition of electrolytes results in destabilization of colloidal contaminants.	Sludge settling rate; dewatering characteristics; bacterial inactivation capacity; sludge stability	Economically feasible; removal of wide variety of pollutants possible	High sludge production	High chemical consumption; large volumes of metallic hydroxides generated as by-product
Adsorption and filtration (activated charcoal, membrane)	Physico-chemical adsorption arising due to complementary surface chemistries of adsorbent and adsorbate	Adsorption capacity; adsorption kinetic constants	Quick removal of charged species	Expensive, especially when non-clay adsorbents are used	regeneration is difficult; highly selective process; significant sludge production when activated carbon is not used
Photochemical (photocatalysis)	Photogeneration of reactive oxidative species on semiconductor-water interface; catalyst loading, pH, temperature, dissolved oxygen, light source (wavelength, intensity) are the key factors.	Photodegradation efficiency; kinetic constants	Low cost, eco-friendly and potentially sustainable; no sludge generation	Formation of byproducts	Scalability and commercialization is often a challenge; catalyst regeneration is non-trivial; thermostatic conditions often essential
Ion-exchange	Reversible exchange of ionic pollutants with benign H^+ and OH^-	Metal rejection rate; removal efficiency	Recyclability; portability	Not effective for all types of pollutants	Suitable ion exchange resins are not easily available for all types of heavy metal species; high operational cost

Reverse osmosis, UV treatment	Applying hydrostatic pressure greater than the osmotic pressure to ensure passage of water, but not ions, across a semiperme-able membrane;	Metal removal efficiency; disinfection efficiency	High salt rejection; resistance to biologi-cal attack; mechanical robustness; chemical stability	Only 35–85% water recovery possible; concentrated elutes haz-ardous for humans, and marine life forms	Costly since membrane porosity, hydrophilicity, thickness, and charge need to be engineered; membrane foul-ing is a major issue

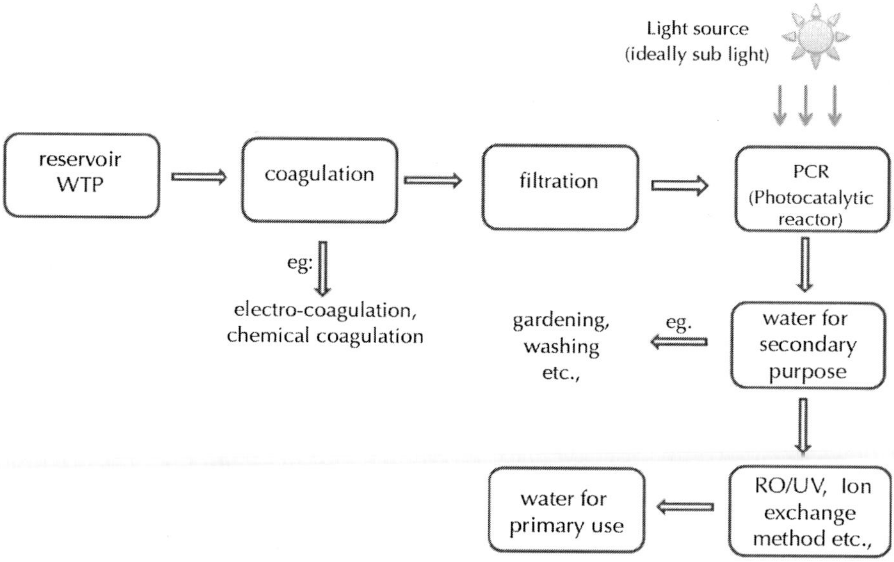

Figure 2: A comprehensive water treatment plant (WTP) will need to employ a multi-step approach to detoxify and sanitize waste water. In case viable photocatalysts are available; coagulation and filtration units can be followed by photocatalytic reactors (PCR). Treated water will be useful for secondary applications. For primary applications (eg. human consumption), RO/UV and ion exchange methods is still likely to be essential. Efficient PCRs are expected to minimize water wastage and energy usage in the post-secondary water treatment process.

ENGINEERING PERSPECTIVES ON PHOTOCATALYST SYNTHESIS AND 'MATERIALS' CHALLENGES

Japan and Germany have industrial associations called Photocatalysis Industry Association of Japan (PIAJ) and German Federation for Applied Photocatalysis (FAP). These organizations develop technology roadmaps for the photocatalytic technologies. Globally photocatalytic water purification faces stiff competition from other mature purification technologies (eg. membrane nanotechnology methods, ultrafiltration, adsorption, chemical or electro-coagulation, distillation etc) (Theresa et al., 2011 and Legrini et al., 1992). Economic viability along with ecological concerns will determine the long term acceptance and impact of photocatalysts. Hence one can legitimately be concerned about recent reports wherein noble metals such as Au and Pt are used for improving photoactivity (Zhao et al., 2013 and Lu et al., 2012). From the point of view of fundamental science, such reports are very appealing; but these technologies would not be economically viable.

For achieving global impact, industries and academics alike must start thinking about environmental impact, and carbon foot print associated with the synthesis and deployment of photocatalysts. Table 2provides an overview of current material classes used for photocatalysis (Zhao et al., 2013, Ren et al., 2012, Zhang et al., 2012a, Palanisamy et al., 2013 and Stender et al., 2008). Synthesis methods, advantages and disadvantages of these materials are tabulated in a manner in which fundamental chemical aspects as well as relevant engineering issues are highlighted. Here the hazard/toxicological information provided comes from the standard materials safety data sheets (MSDS) of representative materials. A multi-disciplinary perspective (Fig. 3) is essential for ensuring that academic photocatalysis research focuses on the most viable technology directions.

Table 2: Material classes used for photocatalysis: examples, synthesis methods, advantages, and disadvantages

Material classes	Examples	Synthesis methods	Advantages	Disadvantages/Remarks
Oxides (Ren et al., 2012)	TiO_2, ZnO, Fe_2O_3, SnO_2, MgO,	Sol–gel, combustion, hydrothermal, solvothermal, soft templating	Easy synthesis and handling, chemically stable, relatively well-understood	None of the oxides (with the exception of TiO_2) have been commercialized; complex photoactivity-synthesis relations
Low dimensional materials (Zhang et al., 2012a)	Carbon nanotubes, graphene, activated carbon	Chemical vapor deposition, spray pyrolysis, arc discharge, laser ablation	High specific surface area, potentially benign	Toxicological data largely unavailable; low synthesis yield; technically challenging synthesis processes involved
Composites (Zhao et al., 2013, Zhang et al., 2012a and Palanisamy et al., 2013)	Graphene–TiO_2,TiO_2–ZnO, ZnO–CuO, TiO_2–SnO_2, TiO_2–Fe_2O_3, plasmonic photocatalysts	Exfoliation method, sol–gel, combustion	Enhanced life times of photoexcited charge carriers; improvement in photoactivity due to electron-hole separation via heterojunctions	Interfacial contact is crucial; often non-benign organic solvents needed for synthesis; usually multi step synthesis required; plasmonic photocatalysts use Au and Pt nanoparticles which make it economically unfeasible
Chalcogenides (Stender et al., 2008)	$NbSe_2$, $TaSe_2$, TaS_2,	Chemical nanofabrication	Fairly good shape and morphology control	Several chalcogenide based compounds such as selenides are hazardous

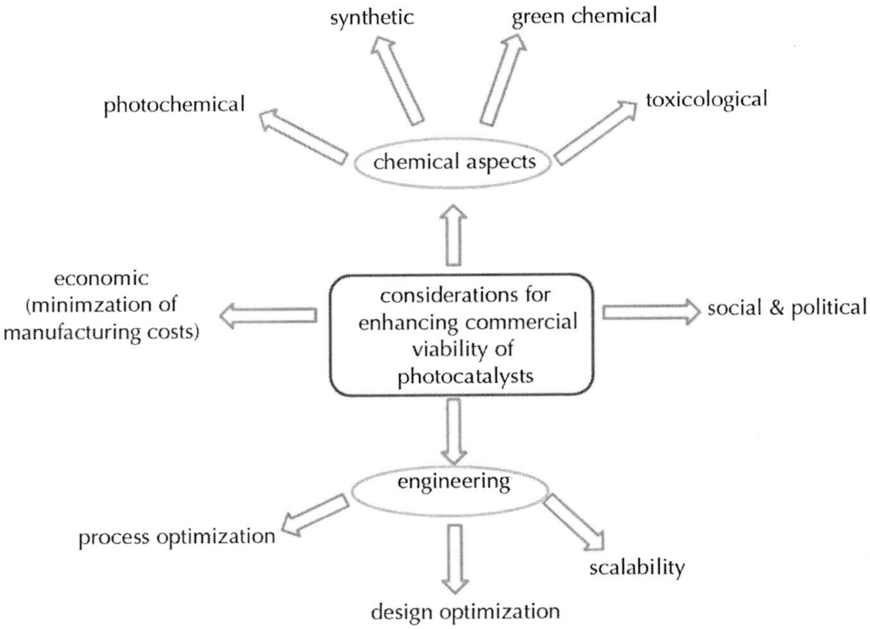

Figure 3: Viability of photocatalysts depends on engineering, economic, chemical and social factors. Hence a multi-pronged engineering approach is essential while designing, synthesizing, and deploying photocatalysts.

Potential nanotoxicity of photocatalysts needs to be considered while developing new material systems. To highlight a particular example, toxicological studies reveal that TiO_2 below 20 nm and ZnO below 30 nm are cytotoxic (Kumar et al., 2011). And yet there exist numerous reports wherein smaller and smaller particles of these systems are being explored for photocatalysis. In particular, the use of chalcogenide quantum dots (QD) for enhancing photocatalytic efficiency (Chaguetmi et al., 2013) may not be industrially viable due to (i) the low yield of QD synthesis, and (ii) unknown toxicological implications of such small particles (typically <10 nm).

Use of rare earths (Zheng et al., 2012) and transition metal ions (Barick et al., 2010) as dopants are known to significantly improve photocatalytic activity. However cytotoxicity associated with heavy metals is an issue.

In some recent reports, Ru-based dyes have been used for photoactivity improvement (especially in the visible region of the solar spectrum). However the scarcity of Ru and its inherent toxicity raises concerns about such directions. At the moment it is not even clear if dye-based sensitization of photocatalyst (Yoon et al., 2010) is a good direction, since the dyes themselves can be potentially harmful. Besides dyes often reduce the recyclability of photocatalysts. Among transition metal dopants, there exist some benign choices (eg. Fe, Mn) (Narayana et al., 2011). One must bear in mind that benign elements can result in harmful chemical complexes. A good example of this is Pd. In its elemental form Pd is quite harmless; however when present in the form of a complex (especially in 2+ state), it can act as an enzyme inhibitor (Liu et al., 1979). From a purely pragmatic point of view, benign choices are better. For example, the recent trend wherein several groups have turned to carbon quantum dots (CQDs) seem promising (Zhang et al., 2012b) due to the benign nature of elemental carbon. Likewise semiconductor composites in which particles are in a carefully chosen size regime, so as to prevent nanotoxicity, is a relatively safe direction (Mary et al., 2013).

Synthesis approaches that require large high-temperature furnaces and sophisticated equipment likely provide road blocks to photocatalysis manufacturing firms. In addition to the infrastructure and energy cost, high temperature processes have a significant carbon footprint associated with them. Hence it makes sense to explore "chimie douce" (soft chemical) approaches (Gopalakrishnan, 1995). Another major determinant of the success of a photocatalyst would be its ability to operate well (i.e. maintain photoactivity) in the presence of co-contaminants (eg. alcohols, humic acids, salts etc.), which are always present in waste water. Very often these co-contaminants drastically quench the activities of the photocatalysts (Uyguner-Demirel and Bekbolet 2011). In order to deal with all these technology challenges, fundamental chemistry will need to meet engineering.

Engineers and scientists working in this water remediation must also be aware of social attitudes that are likely to impact their technology goals. It should come as no surprise that the social acceptance of a technology is intimately connected to the economics involved, and the prevalent legal structures present on-field. Unless benefits associated with alternative remediation technologies is evident, and affordability is ensured (Santosa et al., 2011), there will social resistance to treated fresh water that comes with a cost.

IDENTIFYING DIRECTIONS BASED ON LESSONS FROM HISTORY

The battery materials community has been continuously developing chimie douce approaches (especially template based routes) for over 2 decades now (Gopalakrishnan, 1995 and Sanchez et al., 2010). In stark contrast, the photocatalysis community has paid little attention to this fertile area of chemistry, which lends itself to green technologies. Alkali metal ions as templating structures is relatively unexplored within the context of photocatalysis. In a recent paper, we reported good photoactivities and adsorptive capacities in (Zn,Cu)O nanocomposites made using a room temperature processes (Mary et al., 2013). We suspect the reason for the high activity, despite the low specific surface area of this system, is due to the presence of alkali ions in the photocatalyst synthesis chamber. Hence chances are that benign alkali ions can not only be used as templating/structure directing precursors; they can also be used for modification of surface chemical properties. Hence there exists a rich repository of one-pot reactions (Sanchez et al., 2010), using which industrially viable photocatalysts can be pursued.

CURRENT SITUATION AND PROSPECTIVE FUTURE DIRECTIONS

PIAJ and FAP focus on photocatalysts for applications involving air and water remediation, medicine/cosmetics and construction materials (ref: PIAJ and FAP homepages; Fig. 1). We are not aware of the exact size of the current global market for photocatalysts. Data from 2006 suggests that by then, applied photocatalysis was edging towards becoming a 10 billion dollar industry (Fujishima and Zhang, 2006). To our knowledge trends in this market have not been well studied or reported since 2006. In particular antifouling, anti-soiling, anti-clouding, anti-bacteria, anti-fungus, and self-cleaning properties of photocatalysts (Fig. 1) are being actively explored at the moment. Gauging by the increasing participation of China, Thailand, Vietnam, Mexico, Singapore and Malaysia (ref: PIAJ and FAP pages), the market is certainly on an upswing. However the market potential is not yet fully tapped, since emerging economies like India and Nigeria are largely invisible in this technology scene. Hence much benefit would come out of engineering research that uses a multi-pronged approach to photocatalytic water remediation (Fig. 3).

CONCLUSIONS

Researchers working on applied photocatalysis need to adopt a multidisciplinary approach to ensure maximum social impact. We suggest that researchers in this area take into account economic, toxicological and carbon-footprint related factors while designing and synthesizing photocatalysts. We think this is essential since, at the moment, the biggest bottleneck is our inability to achieve the translation of lab technologies.

Recent directions wherein noble metals or rare earths are being used to modify photocatalytic activities are scientifically interesting, but are likely to be commercially non-viable. Combining chimie douce and green chemical principles is a useful way to achieve industrially viable photocatalysts. The long standing problem of low-cost, energy-efficient water treatment reactor will likely involve an integration of photocatalysis based reactors into the existing process line.

ACKNOWLEDGMENTS

Tiju Thomas acknowledge the Department of Science and Technology for financial support in the form of INSPIRE Faculty Grant (DST 0117). He also thanks Prof. Giridhar Madras (Chemical Engineering, Indian Institute of Science, Bangalore) for many helpful discussions.

REFERENCES

1. Bahnemann, D.W., 2000. Current challenges in photocatalysis: Improved photocatalysts and appropriate photoreactor engineering. Research on Chemical Intermediates 26, 206–220.
2. Barick, K.C., Singh, S., Aslam, M., Bahadur, D., 2010. Porosity and photocatalytic studies of transition metal doped ZnO nanoclusters. Microporous and Mesoporous Materials 134, 195–202.
3. Centre, U.N., 2009a. Ballooning global population adding to water crisis, warns new UN report, United Nations, Geneva.
4. Centre, U.N., 2009b. Majority of world population face water shortages unless action taken. warns Migiro, United Nations, Geneva.
5. Chaguetmi, S., Mammeri, F., Nowak, S., Decorse, P., Lecoq, H., Gaceur, M., Naceur, J.B., Achour, S., Chtourou, R., Ammar, S., 2013. Photocatalytic activity of TiO2 nanofibers sensitized with ZnS quantum dots. RSC Advances 3, 2572–2580.
6. Chan, G.Y.S., Wai-Hung, L., Sandhya, B., 2006. Physico-chemical treatment techniques for wastewater laden with heavy metals, Tonni Agustiono Kurniawan. Chemical Engineering Journal 118, 83–98.
7. Chong, M.N., Jin, B., Chow, C.W.K., Chris, S., 2010. Recent developments in photocatalytic water treatment technology: A review. Water Research 44, 2997–3027.

8. Fujishima, A., Zhang, X., 2006. Titanium dioxide photocatalysis: present situation and future approaches. Comptes Rendus Chimie 9, 750–760.
9. Gopalakrishnan, J., 1995. Chimie Douce approaches to synthesis of metastable oxide materials Chemistry of Materials 7, 1265–1275.
10. Khin, M.M., Sreekumaran Nair, A., Jagadeesh Babu, V., Murugana, R., Ramakrishna, S., 2012. A review on nanomaterials for environmental remediation. Energy and Environmental Science 5, 8075–8109.
11. Kumar, A., Pandey, A.K., Singh, S.S., Shanker, R., Dhawan, A., 2011. Cellular uptake and mutagenic potential of metal oxide nanoparticles in bacterial cells. Chemosphere 83, 1124–1132.
12. Legrini, O., Oliveros, E., Braun, A.M., 1992. Photochemical Processes for Water Treatment. Chemical Reviews 93, 671–698.
13. Liu, T.Z., Lee, S.D., Bhatnagar, R.S., 1979. Toxicity of palladium. Toxicology Letters 4, 469–473
14. Lu, Y., Yu, H., Chen, S., Quan, X., Zhao, H., 2012. Integrating plasmonic nanoparticles with TiO2 photonic crystal for enhancement of visible-light-driven photocatalysis. Environmental Science and Technology 46, 1724–1730.
15. Mary, N., Madras, G., Kottam, N., Thomas, T, 2013. Cu based enhancement of solar spectrum absorbance and photoactivity of ZnO. Journal of Hazardous Materials communicated.
16. McCullagh, C., Skillen, N., Adams, M., Robertson, P.K.J., 2011. Photocatalytic reactors for environmental remediation: a review. Journal of Chemical Technology and Biotechnology 86, 1002–1017.
17. Narayana, R.L., Matheswaran, M., Aziz, A.A., Saravanan, P., 2011. Photocatalytic decolourization of basic green dye by pure and Fe, Co doped TiO2 under daylight illumination. Desalination 269, 249–253.

18. Nations, U, 2008. Tackling a global crisis. International Year of Sanitation, 2008.
19. Palanisamy, B., Babu, C.M., Sundaravel, B., Anandan, S., Murugesan, V., 2013. Sol–gel synthesis of mesoporous mixed Fe2O3/TiO2 photocatalyst: application for degradation of 4-chlorophenol. Journal of Hazardous Materials 252–253, 233–242.
20. Pérez-González, A., Urtiaga, A.M., Ibáñez, R., Ortiz, I., 2012. State of the art and review on the treatment technologies of water reverse osmosis concentrates. Water Research 46, 267–283.
21. Prüss-Üstün, A., Bos, R., Gore, F., Bartram, J., 2008. Safer Water, Better Health: Costs, benefits, and sustainability of interventions to protect and promote health. World Health Organization, Geneva.
22. Ren, Y., Zhen, M., Bruce, P.G., 2012. Ordered mesoporous metal oxides: synthesis and applications. Chemical Society Reviews 41, 4909–4927.
23. Sanchez, C., Rozes, L., Ribot, F., Laberty-Robert, C., Grosso, D., Sassoye, C., Boissiere, C., Nicole, L., 2010. "Chimie douce": a land of opportunities for the designed construction of functional inorganic and hybrid organic-inorganic nanomaterials. Comptes Rendus Chimie 13, 3–39.
24. Santosa, A.C., Roberts, J.A., Barreto, M.L., Cairncross, S., 2011. Demand for sanitation in Salvador. 72. A hybrid choice approach, Social Science & Medicine, Brazil, pp. 1325–1332.
25. Stender, C.L., Sekar, P., Odom, T.W., 2008. Solid-state chemistry on a surface and in a beaker: unconventional routes to transition metal chalcogenide nanomaterials. Journal of Solid State Chemistry 181, 1621–1627.
26. Theresa, M., Pendergast, M., Hoek, E.M.V, 2011. A review of water treatment membrane nanotechnologies. Energy and Environmental Science 4, 1946–1971.

27. Uyguner-Demirel, C.S., Bekbolet, M., 2011. Significance of analytical parameters for the understanding of natural organic matter in relation to photocatalytic oxidation. Chemosphere 84, 1009–1031.
28. UNESCO/WHO, 2009. Diarrhoea: Why children are still dying and what can be done. (http://www.unicef.org/media/files/Final_Diarrhoea_Report_Octo ber_2009_final.pdf).
29. Yoon, T.P., Ischay, M.A., Du, J., 2010. Visible light photocatalysis as a greener approach to photochemical synthesis. Nature Chemistry 2, 527–532.
30. Yu, K., Yang, S., Liu, C., Chen, H., Li, H., Sun, C., Boyd, S.A., 2012. Degradation of organic dyes via bismuth silver oxide initiated direct oxidation coupled with sodium bismuthate based visible light photocatalysis. Environmental Science and Technology 46, 7318–7326.
31. Zhang, H., Huang, H., Ming, H., Li, H., Zhang, L., Liu, Y., Kang, Z., 2012b. Carbon quantum dots/Ag3PO4 complex photocatalysts with enhanced photocatalytic activity and stability under visible light. Journal of Materials Chemistry 22, 10501–10506.
32. Zhang, Y., Fei, L., Jiang, X., Pan, C., Wang, Y., 2011. Engineering nanostructured Bi2WO6–TiO2 toward effective utilization of natural light in photocatalysis. Journal of American Ceramic Society 94, 4157–4161.
33. Zhang, Y., Zhang, N., Zi-Rong, T., Yi-Jun, X., 2012a. Improving the photocatalytic performance of graphene–TiO2 nanocomposites via a combined strategy of decreasing defects of graphene and increasing interfacial contact. Physical Chemistry Chemical Physics 14, 9167–9175.
34. Zhao, S., Ramakrishnan, G., Shen, P., Su, D., Orlov, A., 2013. The first experimental demonstration of beneficial effects of sub-nanometer platinum particles for photocatalysis. Chemical Engineering Journal 217, 266–272.

35. Zheng, Y., Chen, Y., Yang, C., Wang, Q., Lin, J., Zhang, L., 2012. Design of europium doped SiO_2–TiO_2 hybrids as novel luminescent photocatalyst. Journal of Luminescence 132, 1639–1641

Chapter 3

Multistage Treatment System for Raw Leachate from Sanitary Landfill Combining Biological Nitrification–denitrification/Solar Photo-Fenton/Biological Processes, at a Scale Close to Industrial – Biodegradability Enhancement and Evolution Profile of Trace Pollutants

Tânia F.C.V. Silva[a], M. Elisabete F. Silva[b], A. Cristina Cunha-Queda[c], Amélia Fonseca[d], Isabel Saraiva[d], M.A. Sousa[e, f], C. Gonçalves[e], M.F. Alpendurada[e], Rui A.R. Boaventura[a], and Vítor J.P. Vilar[a]

[a]LSRE e Laboratory of Separation and Reaction Engineering, Associate Laboratory LSRE/LCM, Faculdade de Engenharia, Universidade do Porto, Rua Dr. Roberto Frias, 4200-465 Porto, Portugal

[b]Escola Superior de Tecnologia de Viseu, Viseu, Portugal

[c]UIQA, Instituto Superior de Agronomia, Universidade Te´cnica de Lisboa (TULisbon), Portugal

[d]Efacec Engenharia e Sistemas S.A. (Unidade de Nego´cios Ambiente), SA, Rua Eng. Frederico Ulrich e Guardeiras, Apartado 3003, 4471-907 Moreira da Maia, Portugal

[e]IAREN e Water Institute of the Northern Region, Rua Dr. Eduardo Torres 229, 4450-113 Matosinhos, Portugal

[f]Laboratory of Bromatology and Hydrology, Faculty of Pharmacy, University of Porto, Rua de Jorge Viterbo Ferreira 228, 4050-313 Porto, Portugal

ABSTRACT

A multistage treatment system, at a scale close to the industrial, was designed for the treatment of a mature raw landfill leachate, including: a) an activated sludge biological oxidation (ASBO), under aerobic and anoxic conditions; b) a solar photo-Fenton process, enhancing the bio-treated leachate biodegradability, with and without sludge removal after acidification; and c) a final polishing step, with further ASBO.

The raw leachate was characterized by a high concentration of humic substances (HS) (1211 mg C_{HS}/L), representing 39% of the dissolved organic carbon (DOC) content, and a high nitrogen content, mainly in the form of ammonium nitrogen (>3.8 g NH_4^+–N/L).

In the first biological oxidation step, a 95% removal of total nitrogen and a 39% mineralization in terms of DOC were achieved, remaining only the recalcitrant fraction, mainly attributed to HS (57% of DOC). Under aerobic conditions, the highest nitrification rate obtained was 8.2 mg NH_4^+–N/h/g of volatile suspended solids (VSS), and under anoxic conditions, the maximum denitrification rate obtained was 5.8 mg (NO_2^-–N + NO_3^-–N)/h/g VSS, with a C/N consumption ratio of 2.4 mg CH_3OH/mg (NO_2^-–N + NO_3^-–N).

The precipitation of humic acids (37% of HS) after acidification of the bio-treated leachate corresponds to a 96% DOC abatement. The amount of UV energy and H_2O_2 consumption during the photo-Fenton reaction was 30% higher in the experiment without sludge removal and, consequently, the reaction velocity was 30% lower. The phototreatment process led to the depletion of HS >80%, of low-molecular-weight carboxylate anions >70% and other organic micropollutants, thus resulting in a total biodegradability increase of >70%.

The second biological oxidation allowed to obtain a final treated leachate in compliance with legal discharge limits regarding water bodies (with the exception of sulfate ions), considering the experiment without sludge.

Finally, the high efficiency of the overall treatment process was further reinforced by the total removal percentages attained for the identified organic trace contaminants (>90%).

GRAPHICAL ABSTRACT

INTRODUCTION

Although the disposal of solid wastes in sanitary landfills constitutes nowadays the most common method of waste management (Robinson and Maris, 1985; Lema et al., 1988), the generation of leachates is inevitable, mainly due to rainwater percolation through wastes and their decomposition products (Renou et al., 2008), leading to a complex mixture of high-strength organic and inorganic contaminants (Baun et al., 2004; Öman and Junestedt, 2008; Zhang et al., 2009). Different recalcitrant contaminants include personal care products (PCPs), pharmaceuticals, hormones, phthalates, humic and fulvic acids (Fan et al., 2006), PAHs (Jonsson et al., 2003; Marttinen et al., 2003), PBDEs (Osako et al., 2004; Wu et al., 2011), pesticides (Öman and Junestedt, 2008) and heavy metals (Jensen et al., 1999; Christensen et al., 2001).

During the last 10 years, publications regarding sanitary landfill leachates treatment rose continuously, achieving more than 1000 papers in 2011 (Fig. 1). The scientific community research interests have been focused on biological, membrane and advanced oxidation processes (AOPs) technologies for the treatment of leachates (Fig. 1). Leachates are normally characterized by a high non-biodegradable organic fraction. In order to reach the necessary quality level of the final effluent fully reducing leachates› negative impact in the environment, in compliance with discharge regulations and at an affordable price, the best treatment system can be obtained using a combination of different technologies, including: i) aerobic and anaerobic biological processes; ii) chemical and physical methods (flotation, coagulation/flocculation, chemical precipitation, adsorption, ammonium stripping, chemical oxidation and ion exchange); iii) membrane filtration (microfiltration, ultrafiltration, nanofiltration and reverse osmosis); iv) AOPs (TiO_2/UV, H_2O_2/UV, Fenton (Fe^{2+}/H_2O_2)/photo-Fenton (Fe^{2+}/H_2O_2/UV)/electro-Fenton/electro-photo-Fenton, ozone (O_3, O_3/UV, and O_3/H_2O_2), etc.). (Wiszniowski et al.2006; Abbas et al., 2009; Gotvajn et al., 2009).

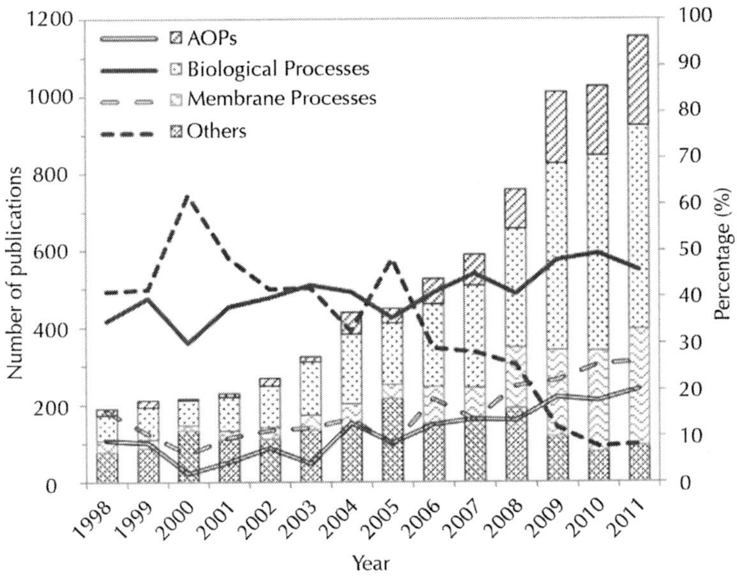

Figure 1: Numbers on research articles concerning landfill leachate treatment technologies per year: SCOPUS search results in the period 1998–2011 (keyword "Landfill Leachate Treatment").Number of publications – bars; Percentage – lines.

Rocha et al. (2011) and Vilar et al. (2012a) applied different photochemical (H_2O_2/UV), heterogeneous (TiO_2/UV, TiO_2/H_2O_2/UV) and homogenous (Fe^{2+}/H_2O_2/UV) photocatalytic processes, including also the Fenton reaction (Fe^{2+}/H_2O_2), to the treatment of real leachates, after preliminary lagooning, from a sanitary landfill located in North Portugal. The photo-Fenton reaction provided degradation rates 20 times higher than the heterogeneous photocatalytic processes. Different other results reported by Vilar et al. (2011a,b,2012a) showed that the solar photo-Fenton oxidation process is an effective technology to improve the biodegradability of mature landfill leachates, allowing to combine AOPs with biological oxidation systems, minimizing chemicals consumption and energy. Silva et al. (2013a) presented the first pre-industrial plant for leachates treatment from a sanitary landfill, installed in the North of Portugal. The treatment strategy consisted on a pre-oxidation of the leachate after lagooning, enhancing its biodegradability and allowing its further coupling with a biological oxidation system, under aerobic and anoxic conditions, to achieve complete nitrogen removal by nitrification and post-denitrification, using methanol as external carbon source. From June 2010 to May 2011, an extensive physico-chemical characterization of the leachate after lagooning showed its high recalcitrant character, mainly associated with the presence of humic substances (HS), which represented almost 59% of the organic carbon content. The acidification of the leachate after lagooning to a pH near 2.8 was a critical point in the photo-Fenton reaction. Silva et al. (2013a) showed that, after acidification, the DOC abatement was approximately 27%, which corresponded to a reduction of HS of ca. 33%, indicating that the precipitation of the humic acids corresponded to approximately 72% of the DOC abatement. Different authors (Zhang et al., 2005; Primo et al., 2008; Wu et al., 2010) reported high mineralization values regarding leachates treatment using Fenton and photo-Fenton processes, although a cautious result analysis must be performed, considering the evaluation of soluble and particulate organic matter content during the treatment. If humic acids precipitation occurs during the acidification process and if the oxidation treatment time is not long enough to degrade those complex molecules present in the particulate phase, after the final effluent neutralization step, those compounds may be redissolved and discharged into water resources.

This work presents a multistage treatment system for raw leachates from a sanitary landfill, at a pre-industrial scale, combining: an activated sludge oxidation process under aerobic and anoxic conditions, promoting nitrification and denitrification reactions in order to achieve a complete removal of nitrogen compounds and the biodegradable organic carbon fraction; followed by a solar photo-Fenton oxidation process (degradation of the most recalcitrant compounds and enhancement of the bio-treated raw leachate biodegradability, without the interference of nitrogen species); and finally an aerobic biological degradation process, for the complete removal of the remaining biodegradable organic compounds. Nitrification and denitrification biological reaction rates were evaluated, as well as alkalinity and methanol consumption, herein used as external carbon source for the denitrification reaction. The solar photo-oxidation efficiency was evaluated considering the sludge removal after acidification and the influence of HS and low-molecular-weight carboxylate anions (LMCA) concentrations in biodegradability enhancement, determined by the Zahn–Wellens test. Finally, some persistent organic micropollutants were identified and their evolution profiles followed-up along the multistage bio-photo-bio-treatment process.

EXPERIMENTAL METHODOLOGY

Chemicals

The photo-Fenton experiments were performed employing hydrogen peroxide (50% (w/v), 1.10 g/cm^3), iron sulfate heptahydrated, and sulfuric acid (98%, 1.84 g/cm^3) for pH adjustment. The photo-treated leachate was neutralized with commercial grade sodium hydroxide (30% (w/v), 1.33 g/cm^3). pH control in the biological reactor was performed with sodium hydroxide (30% (w/v), 1.33 g/cm^3) and sulfuric acid (98%, 1.84 g/cm^3). Methanol (0.79 kg/L) was used as carbon source for biological denitrification reaction. All reactants were of commercial grade provided by Quimitécnica SA. Ultrapure and distilled water were produced by a Millipore system (Direct-Q model) and a reverse osmosis system (Panice®), respectively.

Analytical Determinations

Sulfates, chlorides, nitrates, phosphates, nitrites and carboxylic acids (acetate, propionate, formate, pyruvate, valerate, malonate, maleate, oxalate, phthalate and citrate) were measured by ion chromatography (Dionex ICS-2100) using a Dionex Ionpac (column AS 11-HC 4 × 250 mm; suppressor ASRS®300 4 mm). Sodium, potassium, ammonium, magnesium and calcium were analyzed by ion cromatography (Dionex DX-120), using a Dionex Ionpac (column: CS12A 4 × 250 mm; suppressor: CSRS®300 4 mm). Isocratic elutions were carried out using 30 mM NaOH/20 mM methanesulfonic acid, at a flow rate of 1.5/1.0 mL/min, for anions/cations analyses, respectively. The gradient program used for short-chain organic acids comprised a pre-run for 8 min with 1 mM NaOH, 20 min with 30 mM NaOH and 10 min with 60 mM NaOH, at a flow rate of 1.5 mL/min, using an eluent generator cartrige (Dionex, RFIC™). Dissolved organic carbon (DOC) was measured by NDIR spectrometry, in a TC-TOC-TN analyzer equipped with ASI-V autosampler (Shimadzu, model TOC-V_{CSN}) and calibrated with standard solutions of potassium hydrogen phthalate (total carbon) and a mixture of sodium hydrogen carbonate/sodium carbonate (inorganic carbon). Dissolved nitrogen was measured in the same TC-TOC-TN analyzer, coupled with a TNM-1 unit (Shimadzu, model TOC-V_{CSN}), via thermal decomposition and NO detection by a chemiluminescence method, calibrated with standard solutions of potassium nitrate.

Total suspended solids (TSS), volatile suspended solids (VSS), sludge volumetric index (SVI), chemical oxygen demand (COD) (after proper dilution in order to avoid the interferences of high chloride content) and total phosphorous were measured according to the Standard Methods for the Examination of Water and Wastewater (Clesceri et al., 2005).

H_2O_2 concentration was determined by reaction with ammonium metavanadate and measuring its characteristic absorbance at 450 nm according to Nogueira et al. (2005) Iron concentration was determined by colorimetry with 1,10-phenantroline, according to ISO 6332. Total polyphenols were quantified by spectrometry at 765 nm, using the Folin-Ciocalteau reagent (Merck) (Folin and Ciocalteau, 1927), and expressed as mg/L of caffeic acid. All samples were filtered through 0.45 μm Nylon membrane filters before analysis. Due to the leachate›s absorption at the selected wavelengths, a blank (leachate diluted as for the colorimetric analyses) was always prepared, and the absorbance was measured at the same wavelength for correction. pH and temperature were measured using a pH meter HANNA HI 4522.

The extraction of HS from leachates was performed by the acid–base treatment method, according to the Standard Methods for the Examination of Water and Wastewater (Clesceri et al., 2005). Briefly, leachate samples were filtered through 0.45 μm cellulose membrane filters and acidified to pH 2.0. Acidified leachates were percolated through a XAD-8 (Supelite DAX-8, Supelco) resin column. The preparative cleaning of the resin is described by Thurman and Malcom (1981). After percolation of all the leachate through the adsorption column, HS were eluted from the XAD-8 resin using 0.1 M NaOH, in reverse flow. HS concentration in the eluted solution was measured in terms of mg C_{HS}/L, using a TC-TOC-TN analyzer (Shimadzu, model TOC-V_{CSN}).

Sanitary Landfill Leachate Sample

The raw leachate sample was collected at a municipal solid wastes (MSW) sanitary landfill, nearby Porto.Table 1 presents the main physico-chemical characteristics of the raw leachate used in this work.

Table 1: Physico-chemical characterization of the landfill leachate at different treatment phases

Parameters	RL[f]	LBT1[f]	LAA[f]		LPF[f]		LPFN[f]		LBT2[f]		ELV[e]
			With sludge	Without sludge	With sludge	Without sludge	With sludge	Without sludge	With sludge	Without sludge	
pH	8.1	8.4	2.9	2.9	3.0	3.1	7.0	7.4	8.0	7.9	6–9
T (°C)	26.4	25.2	26.5	24.1	32.5	30.2	32.5	23.4	22.6	21.3	
TSS (mg/L)	126	900	1500	93	800	358	132	11	126	45	60
VSS (mg/L)	94	690	1320	93	525	160	76	2	82	31	
DOC (mg C/L)	2503	1534	1200	1200	518	261	577	248	196	45	
IC (mg/L)	4119	1903	0	0	0	0	5.9	3.2	176	122	–

$CaCO_{3,IC}$[a] (g/L)	17.1	8.0	0	0	0	0	0.025	0.013	0.73	0.51	–
Mineralization in each process[b] (%)	–	39	11	13	–	–	27	38	15	8	
Accumulated mineralization (%)	–	39	50	52	–	–	77	90	92	98	
COD (mg O_2/L)	7426	4864	4041	3720	1243	408	1305	436	467	117	150
BOD_5 (mg O_2/L)	1325	200	–	–	–	–	718	249	14	12	40
BOD_5/COD	0.18	0.04	–	–	–	–	0.55	0.57	0.03	0.08	
Humic substances (mg C_{HS}/L)	977 (1211)[c]	872	550	550	50	26	92	30	91	30	
HS/DOC (%)	39	57	46	46	10	10	16	12	46	67	
HS/COD (%)	13	18	12	15	4	6	7	7	20	26	
Abs 254 nm (diluted 1:25)	–	1.466	1.160	1.030	0.082	0.086	0.111	0.106	–	–	
Dissolved Iron (mg (Fe^{2+} + Fe^{3+})/L)	4.8	7.9	23.3[d]	19.7[d]	30.7	22.9	3.6	0.5	1.1	0.7	
Sulfate (mg SO_4^{2-}/L)	42	6831	–	14,522	–	13,917	14,233	14,079	13,931	13,803	2000
Chloride (mg Cl^-/L)	3549	3370	–	3122	–	3136	3209	3171	3192	3060	
Total Nitrogen (mg N/L)	4080	210	219	182	227	195	221	194	232	8,80	15
Ammoniacal nitrogen (mg NH_4^+–N/L)	3864	23	–	34	–	67	64	53	125	<1	8
Nitrate (mg NO_3^-–N/L)	<1	<1	–	<1	–	<1	<1	<1	<1	<1	11
Nitrite (mg NO_2^-–N/L)	139	<1	–	<1	–	<1	3	<1	2	2	
Total phosphorous (mg P/L)	35	25	–	–	–	–	8	<1	4	<1	10

[a]Alkalinity values considering that at pH less than 8.0 the inorganic carbon is almost in the form of bicarbonates (Sawyer et al., 2003).

[b]Mineralization in each treatment stage relatively to the initial DOC (2503 mg/L).

[c]Humic substances concentration in the raw leachate is 1211 mg C_{HS}/L. The mixture of the raw leachate with the activated sludge resulted in 977 mg C_{HS}/L.

[d]High values due to contamination in reactor.

[e]ELV – Emission Limit Values.

[f]RL – Raw Leachate; LBT1 – Leachate after 1st biological treatment; LAA – Leachate after acidification; LPF – Leachate after Photo-Fenton reaction; LPFN - Leachate after Photo-Fenton reaction and neutralization; LBT2 – Leachate after 2nd biological treatment.

Experimental Set-up

Solar CPC Reactor

Photo-Fenton experiments were carried out under natural sunlight, in a pre-industrial CPC-based plant installed at a sanitary landfill, located in North Portugal. The solar plant consists of a photocatalytic system, constituted by 2 separate parallel rows of 10 and 9 CPC modules (10 tubes/module, 2.08 m^2/module), connected by polypropylene junctions, with a total collector surface of 39.52 m^2, mounted on a fixed platform tilted 41° (local latitude), oriented South. The tubes are made of borosilicate glass (Schott-Duran type 3.3, Germany, cut-off at 280 nm, internal diameter 46.4 mm, length 1500 mm and thickness 1.8 mm) connected by polypropylene junctions. Each of the CPC rows is connected to a cylindrical conic tank (1.6 m diameter, 2 m height, volume 2.5 m^3). The leachate is pumped from the recirculation tank to the CPCs, using a centrifugal pump at a maximum flow rate of 65 L/min, regulated by two flow meters. The plant can be operated in two ways: using the two CPC modules (illuminated volume of 482 L) or using one module of 10 (illuminated volume of 254 L) or 9 (illuminated volume of 228 L). The cylindrical conic tank is equipped with a mechanical stirrer (maximum stirring rate of 76 rpm), a pH sensor and controller (Colberge), a H_2O_2sensor and controller (Grundfos Alldos, Conex DIA-1), three dosing pumps (Colberge; Q_{max} = 13 L/h) for H_2SO_4, H_2O_2 and NaOH addition and their respective containers. The photoreactor is a completely autonomous system, controlled by an abstract computer (Magelis, Schneider Electric), constituted by three electrovalves (Ebro Armaturen, P_{max} = 8 bar), three level rods and three valves. A scheme of the pre-industrial plant can be found at Silva et al. (2013a).

The intensity of solar UV radiation was measured by a global UV radiometer (CUV4, Kipp & Zonen) mounted on the CPC reactors at the same angle, which provided data in terms of incident W_{UV}/m^2. Eq. (1)allows to obtain the amount of accumulated UV energy ($Q_{UV,}n$, kJ/L) received on any surface in the same position with regard to the sun, per unit of volume of water inside the reactor, in the time interval t:

$$Q_{UV,n} = Q_{UV,n-1} + \Delta t_n \overline{UV}_{G,n} \frac{A_r}{1000 \times V_t}; \quad \Delta t_n = t_n - t_{n-1} \tag{1}$$

where t_n is the time corresponding to n-water sample (s), V_t total reactor volume (L), A_r illuminated collector surface area (m^2) and $\overline{UV}_{G,n}$ average solar ultraviolet radiation (W/m^2) measured during the period Δt_n (s).

Biological Oxidation System

The biological oxidation system is composed by a cylindrical flat-bottom tank (1.6 m diameter, 2 m height and volume 3.5 m^3), equipped with a mechanical stirrer (maximum stirring rate of 76 rpm), a pH control unit (Colberge) for pH adjustment using either H_2SO_4 or NaOH dosed by means of two metering pumps (Dosapro Milton Roy), a dissolved oxygen sensor and controller (Colberge), a methanol dosing pump (Colberge, Q_{max} = 13 L/h). Air was supplied by a blower (compressor; Q_{max} = 198 L/min) for maintaining the oxygen in the tank in the selected range, using a ceramic diffuser at bottom of the biological tank. The biological reactor is a completely autonomous system, controlled by an abstract computer (Magelis, Schneider Electric), constituted by one electrovalve (Ebro Armaturen, P_{max} = 8 bar) and four discharge valves.

Experimental Procedure

1st Biological Treatment

The biological reactor was initially colonized using activated sludge from a conventional municipal wastewater treatment plant (WWTP), and small amounts of raw leachate were added during one month, in order to adapt the biomass to the high-strength wastewater. After this preliminary adaption period, the clarified raw leachate (2.5 m^3) was pumped into the biological reactor containing the biomass. The biological reactor was operated under aerobic and anoxic conditions, to promote both nitrogen removal, by nitrification and denitrification reactions, respectively, and the complete oxidation of the biodegradable organic carbon fraction. Furthermore, and taking into account the low carbon/nitrogen ratio of the raw leachate, methanol was used as an external carbon source to promote complete denitrification. pH was controlled in the range between 7.5 and 9.0, along nitrification and denitrification reactions. During nitrification, oxygen concentration was maintained above 0.5 mg O_2/L. Over denitrification reactions, the air blower was turned off (mechanical stirrer remained turned on, in order to avoid sludge settling) and dissolved oxygen concentration was always below the detection limit (<0.05 mg O_2/L).

Solar Phototreatment

After the 1st biological treatment and further wastewater clarification, the bio-treated leachate was pumped into the recirculation tank of the photocatalytic system, to proceed with the photo-Fenton oxidation. A first control sample was taken after 15 min of mechanical stirring, for bio-treated leachate characterization. pH was adjusted to 3.0 through the addition of sulfuric acid, under mechanical stirring, and another sample was taken after pH stabilization. In order to evaluate the influence of sludge removal on the photo-Fenton reaction, the acidified bio-treated leachate was divided in two parts: half the volume was transferred to another tank, followed by sludge settling, and the supernatant was transferred back again to the recirculation tank, to perform the photo-oxidation reaction. In the experiments with and without sludge removal after acidification, 0.9 and 0.8 m^3 of acidified-bio-treated leachate was used, respectively. Iron sulfate salt (80 mg Fe^{2+}/L) and a first dose of H_2O_2 (500 mg/L) was added manually and automatically to the wastewater, respectively, achieving a final pH value of 2.8. The presence of dissolved iron concentration in the leachate (7.9 mg/L) (Table 1) was not taken into account. Following, the acidified-bio-treated leachate was pumped through the CPCs (10 CPCs modules; 20.8 m^2) and the photo-Fenton reaction started. Hydrogen peroxide was controlled online between 100 and 500 mg/L, through the addition of a commercial H_2O_2 (50% (w/v)) solution. Samples were taken at pre-defined times to evaluate the degradation process. Reactions stopped automatically when the consumption of H_2O_2 or the amount of UV radiation (integrated by the UV radiometer) reached the set point. At this moment, the pump was turned off and the bio-photo-treated leachate returned to the recirculation tank by gravity. Finally, the bio-photo-treated leachate was neutralized with NaOH to a pH around 7, under mechanical stirring, leading to iron precipitation, followed by an iron sludge settling period of 3 h.

2nd Biological Treatment

The clarified neutralized photo-bio-treated leachate (0.6 m^3) was pumped into the biological reactor, containing the settled biomass. The biological reactor was operated under the same conditions presented in Section 2.5.1. After oxidation of the biodegradable organic carbon content, and further clarification of the wastewater, the bio-photo-bio-treated leachate was discharged.

Biodegradability Assays

Before biological tests and other analyses involving chemical oxidation, the excess of H_2O_2 present in the samples was removed, using a small volume of a 0.1 g/L solution of catalase (2500 U per mg bovine liver), after pH adjustment to 6.5–7.5. Although the contribution of catalase for DOC and COD values is negligible, a blank solution containing the same amount of catalase added to the leachate samples was prepared using pure water, and the resulting DOC and COD values were discounted in the leachate samples.

A 28 days biodegradability Zahn–Wellens test was performed, according to the EC protocol, Directive 88/303/EEC (EPA, 1996). 250 mL of pre-treated samples, at different photo-Fenton times, without hydrogen peroxide and at neutral pH were added to an open borosilicate glass beaker, magnetically stirred and kept in the dark at 25 °C. Activated sludge from a municipal wastewater treatment plant (WWTP) located in Porto, previously centrifuged, and mineral nutrients (KH_2PO_4, K_2HPO_4, Na_2HPO_4, NH_4Cl, $CaCl_2$, $MgSO_4$ and $FeCl_3$) were added to the samples. Control and blank experiments were prepared using glucose as carbon source, which is highly biodegradable, and pure water, respectively, to which were also added mineral nutrients and activated sludge. The percentage of biodegradation (D_t) was calculated by the following equation:

$$D_t = \left[1 - \frac{C_t - C_B}{C_A - C_{BA}}\right] \times 100 \tag{2}$$

where C_A and C_{BA} are the DOC (mg/L) in the sample and in the blank, measured 3 h after starting the experiment, C_t and C_B are the DOC (mg/L) in the sample and in the blank, measured at the sampling time t. The photo-Fenton pre-treated samples are considered biodegradable when D_t is higher than 70% (EMPA, 1992).

Target and Non-target Screening of Persistent Organic Micropollutants

Leachate samples screening analyses for volatile organic compounds (VOCs), pesticides, phenols, phthalates, polycyclic aromatic hydrocarbons (PAHs) and polychlorinated biphenyls

(PCBs) were performed by solid-phase microextraction, followed by gas chromatography coupled to mass spectrometry (SPME-GC-MS).

Leachate samples collected at different treatment stages were first extracted using a polydimethylsiloxane/divinylbenzene (PDMS-DVB) fiber (60 μm), supplied by Supelco (Bellefonte, PA, USA) for the analysis of semivolatile pollutants, while for the analysis of VOCs a Carboxen/PDMS 75 μm fibre was used. Samples collected at an earlier treatment stage, i.e., prior to the phototreatment step, were first diluted (1:10) in some cases, owing to their dark-brown color and intense turbidity. For subsequent analyses, 10 mL samples were used for COVs extraction (headspace), 18 mL for pesticides, phthalates, PAHs and PCBs (fiber immersion), and 18 mL acidified with sulfuric acid to pH 2 for phenols extraction (fiber immersion). In immersion mode, extractions were conducted over 60 min, at 60 °C and with constant stirring at 250 rpm. Two different GC–MS systems were employed for the overall characterization: i) for pesticides, phenols, phthalates, PAHs and PCBs, a Varian CP 3800 (Walnut Creek, CA, USA) gas chromatograph was used, equipped with a fused-silica capillary column coated with 5% diphenylmethylsiloxane (DPMS), VF-5 MS (30 m × 0.25 mm I.D., 0.25 μm film thickness), coupled to a 4000 ion trap mass spectrometer, from Varian Instruments (Walnut Creek, CA, USA); ii) for VOCs, analyses were carried out in a Varian 3400 CX (Walnut Creek, CA, USA) gas chromatograph, equipped with a BR-624 capillary column (30 m × 0.25 mm I.D., 1.4 μm film thickness), coupled to a Saturn 2000 ion trap mass spectrometer, from Varian Instruments (Sunnyvale, CA, USA). Mass spectrometers were operated in full-scan mode, in the range between 35 and 420 m/z.

Further details on the analytical methods herein employed for target quantitative analysis of the different organic micropollutants can be found in Gonçalves and Alpendurada (2004), Beceiro-González et al. (2007) and Guimarães et al. (2008). For non-target screening analyses, spectra of unidentified chromatographic peaks, not present in the respective blank/negative control chromatogram, were compared against reference spectra included in the US

National Institute of Standards and Technology (NIST) database, after background subtraction.

RESULTS AND DISCUSSION

Raw Leachate Characterization

Table 1 presents the characteristics of the raw leachate from a sanitary landfill, which receives about 446,378 tons per year of solid wastes (85% urban solid wastes and 15% non-toxic industrial wastes) since 1999. The landfill covers 384 km^2 and it is located in northern Portugal, serving approximately 446,378 inhabitants. The sanitary landfill produces in average 100 m^3 leachate per day and has leachate pre-treatment system composed by the following treatment units: (i) an aerated lagoon (15,000 m^3) with pure oxygen addition; (ii) an aerobic and an anoxic biological reactor with activated sludge (750 m^3) connected to a clarifier (27 m^3); (iii) a coagulation/flocculation system with chemical sludge removal; and (iv) a final retention lagoon (3000 m^3). The pre-treated leachate is then transported to a municipal WWTP, by cisterns, where a final treatment is conducted. The raw leachate is collected before the aerated lagoon stage and it presents an intense dark-brown color associated to the high concentration of humic substances (>1211 mg C_{HS}/L), which corresponds to 39% and 13% of DOC (2503 mg C/L) and COD (7426 mg O_2/L), respectively. It has also a high nitrogen concentration (4080 mg N/L), mainly in the form of ammonium (95%) and low values of BOD_5/COD (0.18), indicating a low biodegradability fraction. According to leachate classification proposed by Chian and DeWalle (1976), as high pH values (pH > 7.5), COD values between 4000 and 10,000 mg O_2/L, BOD_5/COD in the range of 0.1–0.3, age higher than 10 years and humic substances concentration higher than 30% in terms of organic compounds, can be classified between intermediate and old leachate. In old mature leachates, the dominant organic fraction is refractory (non-biodegradable), since volatile fat acids are converted to biogas (CH_4, CO_2) by the methanogenic microorganisms (Chian and DeWalle, 1976). According to the leachates characteristics, the treatment strategy adopted in this work consists in a multistage treatment system with three sequential steps: preliminary biological treatment by activated sludge under aerobic and anoxic conditions promoting the removal of the biodegradable organic carbon fraction and nitrogen; a solar photo-Fenton process to oxidize the recalcitrant organic compounds into more biodegradable ones; being possible to couple with a second aerated activated sludge process, to achieve complete removal of residual biodegradable organic carbon and nitrogen.

1st Biological Oxidation

Fig. 2 presents the evolution of dissolved organic carbon and nitrogen content during the 1st biological oxidation process. Fast biodegradable organic carbon fraction was totally removed after 24 h, achieving 22% mineralization, with a kinetic constant of 19 mg DOC/h/g VSS. The slow biodegradable organic fraction was almost totally removed, according to the BOD_5 final value achieved, with a kinetic constant of 1.8 mg DOC/h/g VSS, leading to a final mineralization of 39%. The remaining organic carbon fraction can be considered recalcitrant, mainly attributed to humic substances, which represented 57% of DOC at the end of the 1st biological treatment. The slight decrease in humic substances concentration, approximately 11%, can be attributed mainly to adsorption in the activated sludge. The SVI obtained during 1st biological treatment was 96 mL/g, indicating a good sludge settleability, normally below 100 mL/g (Metcalf and Eddy, 2005). Regarding the food to microorganism (biomass) ratio (F/M), it was achieved an acceptable value of 0.08 g substrate/g biomass/day, since typical values for the BOD F/M ratio reported in literature vary from 0.04 g/g/day, for extended aeration processes, to 1.0 g/g/day, for high rate processes (Metcalf and Eddy, 2005).

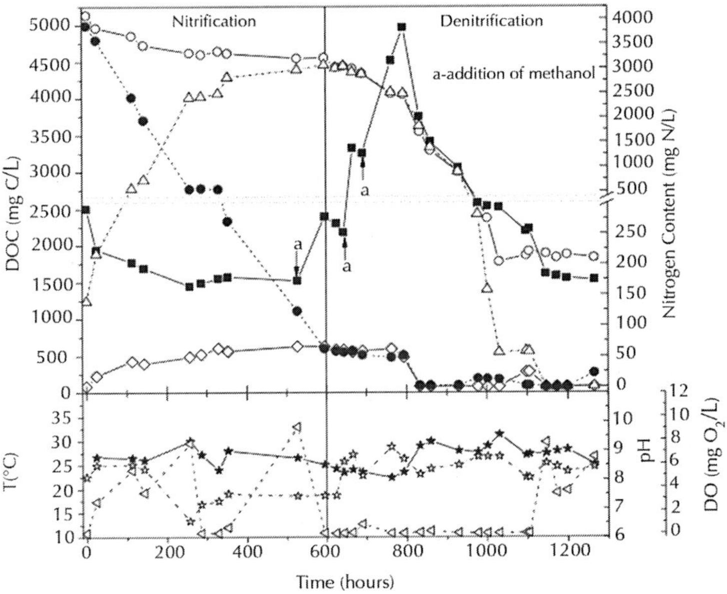

Figure 2: Biological nitrification/denitrification of the raw leachate. ■ – DOC; ○ – Total Nitrogen; ● – Ammonium (NH_4^+–N); ◇ – Nitrate (NO_3^-–N); △ – Nitrite (NO_2^-–N); ★ – T; ☆ – pH; ◁ – Dissolved Oxygen (DO).

Biological nitrogen removal requires a two-step process: aerobic nitrification of ammonia to nitrite (Eq. (2)) and then nitrite is converted to nitrate (Eq. (3)); anoxic denitrification of nitrate/nitrite to nitrogen gas (Eq.(4)).

$$NH_4^+ + \frac{3}{2}O_2 \rightarrow NO_2^- + H_2O + 2H^+ \text{(nitritation, ammonia oxidizing bacteria)} \quad (3)$$

$$NO_2^- + \frac{1}{2}O_2 \rightarrow NO_3^- \text{(nitratation, nitrite oxidizing bacteria)} \quad (4)$$

$$NO_3^- \rightarrow NO_2^- \rightarrow NO \rightarrow N_2O \rightarrow N_2 \text{(denitrification, denitrifying bacteria)} \quad (5)$$

Fig. 2 shows that under aerobic conditions, it was accumulated more than 96% of the loaded ammonium nitrogen as nitrite, with 98% ammonium conversion, achieving final values of 3 g NO_2^-–N/L and 60 mg NO_3^-–N/L. pH was maintained between 7.5 and 8.5 through the addition of NaOH, as reported by Alleman (1984) which showed that the optimal pH values are between 7.9 and 8.2 for nitrification.

Nitrification reaction stopped between 287 and 353 h due to a problem in the base dosing pump, achieving a pH value of 6.6 and also a problem with the compressor, resulting in very low dissolved oxygen concentration values (<0.1 mg O_2/L). The highest nitrification rate obtained was 8.2 mg NH_4^+–N/h/g VSS (T = 26.9 °C) consuming 4.5 g $CaCO_3$ per liter of raw leachate or 1.2 mg $CaCO_3$ per mg NH_4^+–N, which is 6 times lower than the stoichiometric ratio, 7.14 mg $CaCO_3$ per mg NH_4^+–N (Eq. (2)). Raw leachate presents initially an inorganic carbon concentration of 4119 mg C/L, which corresponds to 20.9 g HCO_3^-/L or 17.1 g $CaCO_3$/L, considering that at pH = 8.0, inorganic carbon is almost in the form of bicarbonates (Sawyer et al., 2003). Until 143 h of nitrification, 2205 mg/L (11.2 g HCO_3^-/L or 9.2 g $CaCO_3$/L) of inorganic carbon was consumed, and 1927 mg NH_4^+–N/L was oxidized, leading to an alkalinity consumption of 4.8 mg $CaCO_3$ per mg NH_4^+–N. At 115 h, pH was 6.6, inorganic carbon concentration was negligible, and consequently, nitrification stopped.

Spagni and Marsili-Libelli (2009) reported nitrification rates between 4.9 and 12.6 mg N/h/g VSS (T = 20 °C) for a leachate with 1199 mg NH_4^+–N/L (COD = 2055 mg O_2/L), achieving nitrification and COD removal efficiencies in average of 98% and 20%, respectively. Ruiz et al. (2003) showed that setting DO concentration at 0.7 mg/L, it was possible to accumulate more than 65% of the loaded ammonia nitrogen as nitrite with 98% ammonia conversion (initial ammonia concentration of 610 mg NH_4^+–N/L). Below 0.5 mg/L of DO, ammonia was accumulated and over a DO of 1.7 mg/L complete nitrification to nitrate was achieved. In our case, despite DO concentration was in average 3.2 mg/L, the nitration process (nitrite oxidation to nitrate) was almost negligible. This can be attributed to the high pH values, possible free ammonia in the reactor, which favored the ammonia oxidizing bacteria and inhibited the nitrite-oxidizing organisms, as reported by Canziani et al. (2006), Villaverde et al. (1997) and Suthersan and Ganczarcczyk (1986).

Complete nitrification requires 2 mol of oxygen per mol of ammonia to be nitrified (Eqs. (2) and (3)). If denitrification is to be considered after a nitrification process, partial nitrification to nitrite implies 25% less oxygen demand compared to complete nitrification, and this shortcut of the nitrate would mean a reduction in the total carbon source required for denitrification, because carbon is needed for conversion of nitrate to nitrite, which can yield up to 40% savings in methanol consumption (Eqs. (5) and (6)).

$$6NO_3^- + 5CH_3OH + CO_2 \rightarrow 3N_2 + 6HCO_3^- + 7H_2O(\text{nitrate removal process}) \tag{6}$$

$$6NO_2^- + 3CH_3OH + CO_2 \rightarrow 3N_2 + 6HCO_3^- + 3H_2O(\text{nitrite removal process}) \tag{7}$$

After 25 days, oxygen supply was turned off, methanol was added as external carbon source for the denitrification process and pH was controlled between 8.0 and 9.0, through the addition of H_2SO_4. After a preliminary denitrification lag phase period of 4 days, characterized by a low denitrification rate of 1.3 mg (NO_2^-–N + NO_3^-–N)/h/g VSS, attributed to the denitrifiers bacteria adaptation to the high nitrite concentration, maximum denitrification rate of 5.8 mg (NO_2^-–N + NO_3^-–N)/h/g VSS was observed, with a C/N consumption ratio of 2.4 mg CH_3OH/mg (NO_2^-–N + NO_3^-–N) (7.4 g/9.4 mL of commercial methanol per liter of leachate).

Between 750 and 850 h, an increase of pH until 9.3 (due to a problem in the dosing pump), resulted in a complete conversion of ammonia from the ionic form, NH_4^+, to the molecular form, NH_3, and combined with the release of high amounts of nitrogen gas due to denitrification reaction, lead to NH_3 stripping, and flotation of activated sludge.

During denitrification it was produced 5.3 g of alkalinity, as $CaCO_3$, per liter of leachate, according to the acid addition, which is equivalent to 1.7 g $CaCO_3$ per g of (NO_2^-–N + NO_3^-–N) reduced. According to Eq.(5), during denitrification it is produced 3.57 g (as $CaCO_3$) of alkalinity per g of NO_2^-–N or NO_3^-–N reduced, which is 2 times higher than the obtained value. During denitrification the inorganic carbon concentration increased from 0 to 1.9 g C/L, corresponding to 9.7 g HCO_3^-/L or 8.0 g $CaCO_3$/L, equivalent to 2.6 g $CaCO_3$/g (NO_2^-–N + NO_3^-–N) oxidized. So, the overall alkalinity produced was 4.3 g $CaCO_3$/g (NO_2^-–N + NO_3^-–N) oxidized, which is very near to stoichiometric reaction.

Different methanol demands, as external carbon source, for denitrification process of landfill leachates have been reported: 2.43 g CH_3OH/g NO_3^-–N (3.6 g COD/g NO_3^-–N) (Kulikowska and Klimiuk, 2004), 2.8–3.0 g CH_3OH/g NO_3^-–N (4.5–4.1 g COD/g NO_3^-–N) (Christensson et al., 1994), which is more than 1.3 times higher than the stoichiometric mass ratio between consumed methanol and nitrate N (1.90 g CH_3OH/g NO_3^-–N). However, when nitrite is the predominant specie in solution, methanol consumption during denitrification is approximately 40% less (1.15 g CH_3OH/g NO_2^-–N). Comparing with the values obtained in our study, methanol consumption was more than two times higher when compared with the stoichiometric mass ratio.

Modin et al. (2007) studied the denitrification using methane as external carbon source, showing a C/N ratio of 7.1 g C–CH_4/g

NO_3^-–N, approximately 13 times higher than the stoichiometric mass ratio, according to Eq. (7) (Randall et al., 1992).

$$8NO_3^- + 5CH_4 \rightarrow 4N_2 + 8HO^- + 6H_2O + 5CO_2 \tag{8}$$

Considering a continuous biological treatment system, with a flow rate of 100 m^3/day, it would be necessary an anoxic lagoon (560 m^3) followed by an aerobic lagoon (500 m^3), with a retention time of 5.6 days and 5.0 days, respectively, to achieve complete denitrification of 3.1 kg NO_2^-–N/m^3 and nitrification of 3.9 kg NH_4^+–N/m^3, considering an average VSS of 4 g/L. In order to minimize the consumption of methanol and alkalinity, the treatment sequence can be anoxic followed by an aerobic reactor, with recirculation to the anoxic reactor: a) to take advantage of the biodegradable organic fraction of the raw leachate, which corresponds to 54% of the organic carbon necessary for denitrification, being necessary only 3.4 kg CH_3OH/m^3, b) to take advantage of the raw leachate alkalinity (17.1 kg $CaCO_3$/m^3), and the alkalinity produced during denitrification (13.3 kg $CaCO_3$/m^3), which is sufficient to achieve complete nitrification of 4 kg NH_4^+–N/m^3 (19.2 kg $CaCO_3$/m^3).

Solar Photo-Fenton Oxidation

After almost complete removal of nitrogen and biodegradable organic carbon fraction, and further clarification of the wastewater, the bio-treated leachate was pumped to the recirculation tank of the photoreactor system, and was acidified until pH 3.0, through the addition of sulfuric acid (3.0 L H_2SO_4/m^3). The bio-treated leachate presented 8.0 g alkalinity in terms of $CaCO_3$ (considering that at pH 8.0, alkalinity is mainly due to bicarbonates), and the amount of sulfuric acid needed to neutralize alkalinity corresponds to 77% of the total acid used during acidification.

However, considering that in a full scale plant, the alkalinity of raw leachate and that produced during denitrification, can be used for nitrification, achieving a final bio-treated leachate with a very low alkalinity, which will reduce substantially the amount of sulfuric acid necessary to perform acidification, decreasing sulfate ion concentration, resulting in a substantially increase of photo-Fenton reaction efficiency due to an higher fraction of more photoactive ferric water complexes than ferric sulfate complexes.

After acidification, it was observed an increase of 67% and 91% in the TSS and VSS, respectively, correlated with humic acids precipitation, leading to humic substances and DOC concentration abatement of 37% and 22%, respectively. This indicates that precipitation of the humic acids (37% of the humic substances), corresponds to approximately 96% of DOC abatement. Silva et al. (2013a) performed an HS analysis to the acid sludge resulted from the acidification step, and results revealed a concentration of 130 mg C_{HS} per gram of dry sludge, which corresponded to production of 2.7 kg of HS sludge per cubic meter of leachate. Considering the same concentration of HS in the sludge, 2.5 kg of HS sludge per cubic meter of leachate were produced according to results observed in this work.

In order to assess the influence of sludge produced after acidification in the photo-Fenton reaction, it was performed two experiments, with and without sludge removal. The photo-Fenton reactions were conducted at an initial pH of 2.8, achieved after the addition of iron and the first dose of H_2O_2 (500 mg/L), and with an initial iron concentration of 80 mg Fe^{2+}/L, reported in a previous work (Vilar et al., 2012b), as the optimum iron concentration for this kind of leachate and photoreactor. This pH value was selected for the photo-Fenton reaction because it avoids iron precipitation (Pignatello et al., 2006). During the reaction, H_2O_2 was supplied in multiple small additions, maintaining the H_2O_2 concentration between 100 and 500 mg H_2O_2/L, which has been reported to improve the oxidation rate, avoiding that hydrogen peroxide can be rate-limiting if applied in concentrations that are too low and, minimizing the consumption of H_2O_2 per amount of COD oxidized, considering that high concentration of hydrogen peroxide can compete with contaminants for the hydroxyl radicals generated and also self-decompose into oxygen and water (Bacardit et al., 2007; Prieto-Rodríguez et al., 2011).

Observing the overall photo-Fenton process (see Tables 1 and 2), for the experiment without sludge removal, it was consumed 54.5 kJ/L of accumulated UV energy and 250 mM of H_2O_2, corresponding to a mineralization of 52% and a specific consumption of 13.6 mg of H_2O_2 per mg of oxidized DOC.While, for the experiment with sludge removal, it was consumed a little more amount of accumulated UV energy and H_2O_2, equivalent to 80.4 kJ/L and 265 mM, respectively.However, the mineralization value was higher, 79%, and the consumed H_2O_2 concentration per unit of DOC was lesser, corresponding to 9.5 mg of H_2O_2per mg of oxidized DOC. The mass and volumetric treatment factors (T_{fm} and T_{fv}), defined as the amount of organics substances or volume of contaminated water, respectively, that the system is able to treat per unit of time and surface of solar collectors (Gálvez and Rodríguez, 2003), were 1.17 g C/h/m^2 and 1.72 L/h/m^{2}for the experiment without acid sludge removal, and 0.95 g C/h/m^2 and 1.01 L/h/m^2 for the experiment with acid sludge removal.

Table 2: Process variables as performance indicators

1st Biological treatment			**Photo-Fenton (PF)**		
Parameter			**Parameter**	**With sludge**	**Without sludge**
	VSS (mg/L)	1581	pH_m	2.8	2.9
	TSS (mg/L)	2257	T_m (°C)	40.5	35.2
	SVI (mL/g)	96	Fe_m	37.6	33.0
	k_{0-25h} [a] (mg DOC/h/g VSS)	19	t_{PF} (h)	25.1	38.0
	$k_{25-258h}$ [b] (mg DOC/h/g VSS)	1.8	I_{UV} (W/m^2)	26.1	22.6

Nitrification	$pH_{n,m}$	7.6	Q_{UV} (kJ/L)	54.5	80.4
	$T_{n,m}$	26.9	$[H_2SO_4]$ (mM)	104	104
	DO_m (mg/L)	3.2	$[H_2O_2]$cons. (mM)	250	265
	$t_{BT,}n$ (h)	600	H_2O_2/DOC_{ox} [f] (mg H_2O_2/mg DOC_{ox})	13.6	9.5
	kn (mg NH_4^+–N/h/g SSV)	8.2	k_1 [c] (L/kJ)	0.020 ± 0.006	0.016 ± 0.001
	[NaOH] (mM)/$CaCO_3$ (g/L)	90/4.5	r_0 [d] (mg/kJ)	20 ± 6	12 ± 1
Denitrification	$pH_{d,m}$	8.4	k_{H2O2} [e] (mmol/kJ)	4.6 ± 1.1	3.2 ± 0.2
	$T_{d,m}$	26.4	T_{fm} [f] (g C/h/m²)	1.18	0.95
	$t_{BT,d}$ (h)	430	T_{fv} [g] (L/h/m²)	1.72	1.01
	k_d (mg (NO_3^-–N + NO_2^-–N)/h/g SSV)	5.8			
	C/N (mg CH_3OH/mg (NO_3^-–N + NO_2—N))	2.4			
	$[CH_3OH]$ (mM)	231			
	$[H_2SO_4]$ (mM)/$CaCO_3$ (g/L)	53/5.3			

[a]DOC removal rate between 0 and 25 h.

[b]DOC removal rate between 25 and 258 h.

[c]Pseudo-first-order kinetic constant for DOC degradation (Q_{UV} > 10 kJ_{UV}/L).

[d]Initial DOC reaction rate (Q_{UV} > 10 kJ_{UV}/L).

[e]H_2O_2 consumption rate (Q_{UV} > 10 kJ_{UV}/L).

[f]Mass treatment factor ((DOC_i – DOC_f) × V_t/(t_{PF} × A_r)).

[g]Volumetric treatment factor (V_t/(t_{PF} × A_r)).

Fig. 3 shows that for the same H_2O_2 dose (220 mM), at similar average reaction temperature and average solar UV power, the mineralization efficiency was 80% and 65% for the experiments with and without sludge removal, considering the dissolved organic carbon. The amount of UV energy and H_2O_2 consumed is 30% higher for the experiment without sludge removal, and consequently, reaction velocity is 30% smaller (see Table 2). The presence of suspended solids decreases light penetration, competing as photons absorbers with H_2O_2 and iron species, being necessary a higher reaction time and consequently a higher energy consumption, and also will increase energy and H_2O_2 consumption to achieve the same mineralization of dissolved organic matter, associated to the degradation of particulate organic matter, as it can be seen by the reduction of more than 50% in TSS and VSS (Table 1).

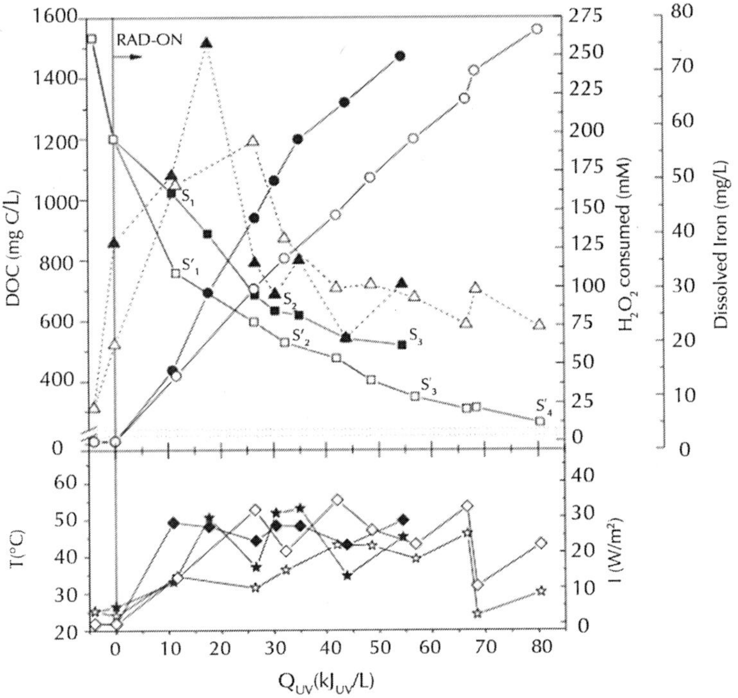

Figure 3: DOC, H_2O_2 consumption, dissolved iron concentration, temperature and average radiation intensity evolution as a function of the accumulated UV energy per liter of leachate during the photo-Fenton process (pH = 2.8; $[Fe^{2+}]$ = 80 mg/L). (■,□) – DOC; (●,○) – H_2O_2 consumed; (▲, △) – Dissolved Iron; (★,☆) – T; (♦,◊) – I; Solid symbols: without sludge removal. Open symbols: with sludge removal.

After photo-bio-treated leachate neutralization, consuming 1.0 kg $CaCO_3$ of alkalinity per cubic meter of leachate, the concentration of humic substances remained approximately constant for experiment without sludge, and increased approximately 184%

for experiment without sludge removal. This indicates dissolution of humic acids present in the particulate phase, which were not totally destroyed during the photo-oxidation, and consequently will be discharged to the receiving water bodies, without treatment, as there are recalcitrant to biological oxidation.

Complete nitrogen removal was not achieved during the first biological treatment, mainly due to organic nitrogen. During the photo-Fenton oxidation it can be observed the oxidation of the organic nitrogen to ammonium nitrogen, which increased from 22.5 to 67.3 mg NH_4^+–N/L (Fig. 4). However, total nitrogen remained approximately constant during the phototreatment.

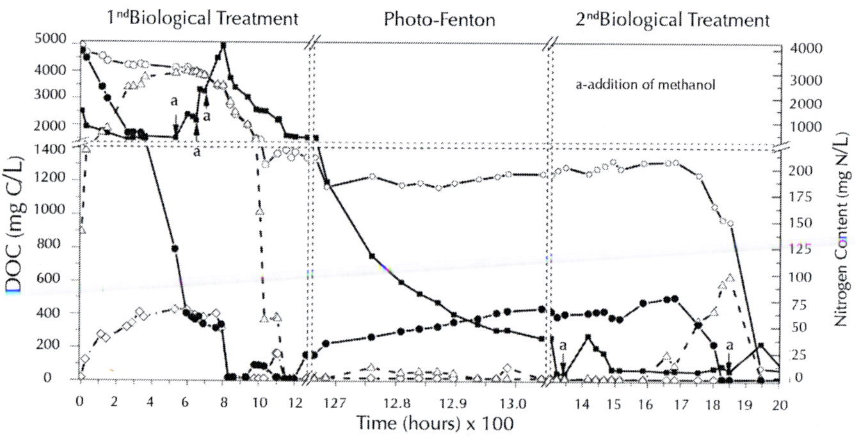

Figure 4: Biological/photo-Fenton/Biological treatment sequence of the raw leachate. ■ – DOC; ○ – Total Nitrogen; ● – Ammonium (NH_4^+–N); ◊ – Nitrate (NO_3^-–N); ▵ – Nitrite (NO_2^-–N).

With the purpose to assess the leachate's biodegradability variation in the course of the treatment process, a Zahn–Wellens test was performed on samples collected at different treatment stages, for both the experiments analyzed so far – with and without sludge removal (Figs. 5 and 6). First samples were collected before and after the 1st biological oxidation process (raw leachate – RL and LBT1, respectively), as well as after the acidification point (LAA). These samples were coincident for both experiments. The remaining samples were taken during the photo-Fenton reaction, after certain doses of H_2O_2 and UV energy (see Fig. 6), for the experiments executed without (S_1, S_2 and S_3) and with (S_1', S_2', S_3' and S_4') removal of the acidic sludge produced during the acidification step.

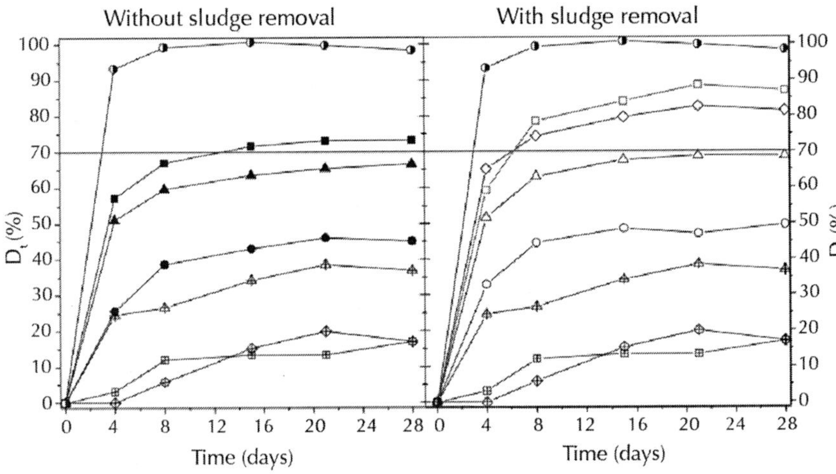

Figure 5: Zahn–Wellens test for samples collected before and after the 1st biological treatment and after acidification (cross symbols), and for some samples taken during the photo-Fenton process without (solid symbols) and with (open symbols) sludge removal after acidification: △ – Raw Leachate; ⊞ – LBT1; ⊕ – LAA; ● – S_1; ▲ – S_2; ■ – S_3; ○ – S_1'; △ – S_2'; ◇ – S_3'; □ – S_4'; ◑ – Reference.

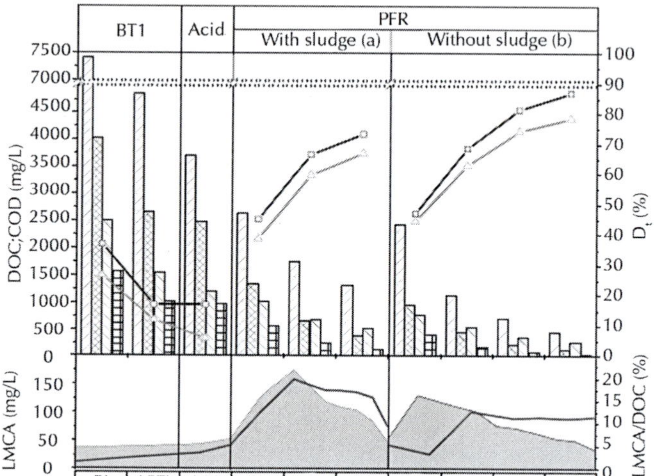

	RL	LBTI	LAA	S1	S2	S3	S'1	S'2	S'3	S'4
Q_{UV}(KJ_{UV}/L)	0	0	0	11.0	27.0	54.0	12.0	32.0	57.0	80.0
H_2O_2(mM)	0	0	0	46	145	250	43	119	196	266
COD(day0)	7426	4864	3720	2640	1755	1305	2440	1130	695	436
COD(day28)	4020	2670	2478	1341	642	378	957	447	207	129
DOC(day0)	2503	1534	1200	1021	684	577	757	528	346	248
DOC(day28)	1573	1013	967	562	232	129	404	161	73	43
D_t(day28)	36.8	17.0	17.0	45.1	66.6	73.3	49.5	68.6	81.4	86.9
D_t(day8)	26.6	12.0	5.8	38.6	59.7	67.0	44.5	62.9	74.3	78.5
LMCA	37.1	40.2	44.2	132	119	48.6	130	101	63.3	33.6
LMCA/DOC	1.5	2.62	3.5	12.9	17.4	9.4	12.7	11.4	10.3	5.8

Figure 6: Evaluation of: DOC and COD during the Zahn–Wellens test at day 0 and day 28; percentage of biodegradation (D_t) during Zahn–Wellens test at day 8 and day 28; low-molecular-weight carboxylate anions (LMCA) concentration and LMCA/DOC ratio, during combined system 1st biological treatment (BT1)/photo-Fenton reaction (PFR), without (a) and with (b) sludge removal after acidification (Acid.).

According to the Zahn–Wellens test, the raw leachate presented a biodegradability percentage of 27% and 37%, after 8 and 28 days, respectively, which is in agreement with the mineralization results obtained in the 1st biological treatment, easily (22%) and slowly (39%) biodegradable organic fraction. Furthermore, COD and DOC values obtained at the end of the 1st biological oxidation were rather close to those achieved by the Zahn–Wellens test. Regarding samples collected before and after acidification, these presented a poor biodegradability percentage (17%), contrary to what happened during the photo-Fenton reaction. Herein, as could be expected (Pulgarin et al., 1999; Wiszniowski et al., 2004), it is observed a biodegradability enhancement; concomitantly with a decrease of HS concentration (95% and 83%, for experiments without and with sludge, respectively). The achieved D_t values were higher than 70%, after only 8 and 15 days, corresponding to a H_2O_2 consumption of 196–266 and 250 mM, for experiments with and without sludge removal, respectively. After a quick analysis of Fig. 5, the treatment stage corresponding to sample S'_3could look like a good endpoint for the pre-oxidation process in the experiment without sludge, given that biodegradability was already higher than 70%. However, if the final purpose of the treatment is the discharge of the treated leachate into water bodies, it would certainly be a better choice to stop the photoreaction in a point close to sample S_4', since then the DOC content is low enough to achieve a COD below 150 mg/L after biological oxidation. These results are in agreement with others obtained in previous work (Silva et al., 2013b), where it was demonstrated that for DOC values lower than 300 mg/L at the end of the phototreatment, it is possible to achieve COD values below 150 mg/L (discharge limit in water bodies) in a subsequent biological treatment.

As aforementioned, Fig. 6 presents the Zahn–Wellens results, at day 0 and day 28, for some samples collected in specific treatment points. Furthermore, this figure also displays the concentration evolution profile of LMCA during the whole photo-Fenton reaction, as well as LMCA/DOC ratios. As predicted, the concentration of total LMCA showed an initial increase followed by a strong decrease, suggesting there was first the breakdown of recalcitrant macromolecules into short-chain carboxylic acids, followed by their degradation (Oturan et al., 2008). Fig. 6 shows maximum LMCA concentration values between 175 and 130 mg C/L (formate > oxalate > phthalate > malonate > acetate > pyruvate > valerate > citrate), corresponding to 13–20% of DOC and a biodegradability percentage higher than 50%, for H_2O_2 doses between 43 and 145 mM.

2nd Biological Oxidation

The neutralized photo-bio-treated leachate suffered a final biological oxidation, resulting in a complete elimination of biodegradable organic carbon fraction (82%) and nitrogen content (95%), regarding the experiment without acid sludge (Fig. 4). In order to perform the last nitrification it was necessary again a long period, mainly attributed to the loss of biomass in the first denitrification step, associated to the foam produced in result of the nitrogen gas release.

Table 1 presents the characteristics of the bio-photo-bio-treated leachate resulted from the two experiments with and without sludge removal. Humic substances concentration didn't suffer any modification during the 2nd biological treatment, representing 46% and 67% of the final DOC, respectively for the experiments without and with sludge removal after acidification. In both experiments, the biological reactor operated with values of SVI close to 100 mL/g and BOD F/M ratio equivalent to 0.2 g/g/day, and it was achieved values of COD below 500 mg O_2/L (DOC < 196 mg C/L), which is in agreement with the discharge limits into the sewer. However, to obey the more stringent regulations related to effluent discharge into receiving water bodies, COD < 150 mg O_2/L, only the experiment performed with sludge removal after acidification reached those requirements, with the exception of sulfate, whose concentration greatly exceeded the discharge limits.

The COD and DOC values obtained after the 2nd biological treatment, for samples S_3 and S_4' (experiments without and with acid sludge removal, respectively), are in accordance with the results of the Zahn–Wellens test (Fig. 6).

Organic Trace Contaminants Identification and Evolution Profile

Using the analytical GC–MS methods described in Section 2.7, an initial target screening analysis was conducted on the raw leachate, as well as on other samples collected in key points of the treatment process, in order to identify and assess the evolution profile of potential trace organic contaminants. Standard solutions were used as reference for the identification and quantification of several VOCs, pesticides, phenols, phthalates, PAHs and PCBs, commonly present in sanitary landfill leachates. However, within the 84 target compounds, only 13 were actually identified and quantified, among which 3 VOCs (benzene, 1,2-dichloroethane and 1,2-dichloropropane), 4 phenols (2-nitrophenol, 2,4-dimethylphenol, 4-chloro-3-methylphenol and p-tert-octylphenol) and 6 PAHs (naphthalene, fluorene, phenanthrene, fluoranthene, pyrene and dibenzo[a,h] anthracene) (Table S1, supplementary material). Therefore, a non-target screening analysis followed, aiming at the identification of other distinct chromatographic peaks, obtained through the four analytical methods.

Fig. 7 presents the GC–MS total ion chromatograms obtained for raw leachate samples (RL), leachate after the 1st biological treatment (LBT1), leachate after the photo-Fenton reaction and subsequent neutralization (LPFN) and the final effluent obtained after the 2nd biological treatment (LBT2). In this way, it was possible to evaluate the individual contribution of each biological treatment (1st biotreatment: RL-LBT1; 2nd biotreatment: LPFN-LBT2) and photo-Fenton process (phototreatment: LBT1-LPFN), for the removal and/or formation of each identified contaminant. Table 3 displays the correspondence between each peak, identified by spectrum comparison against reference spectra from NIST library, and the respective compound characteristics. Simply by visual inspection, it is easily inferred the high efficiency of the overall treatment process, comparing the RL and BLBT2 chromatograms (Fig. 7), as well as by the >90% total removal efficiencies calculated for the identified contaminants, presented in Table 3.

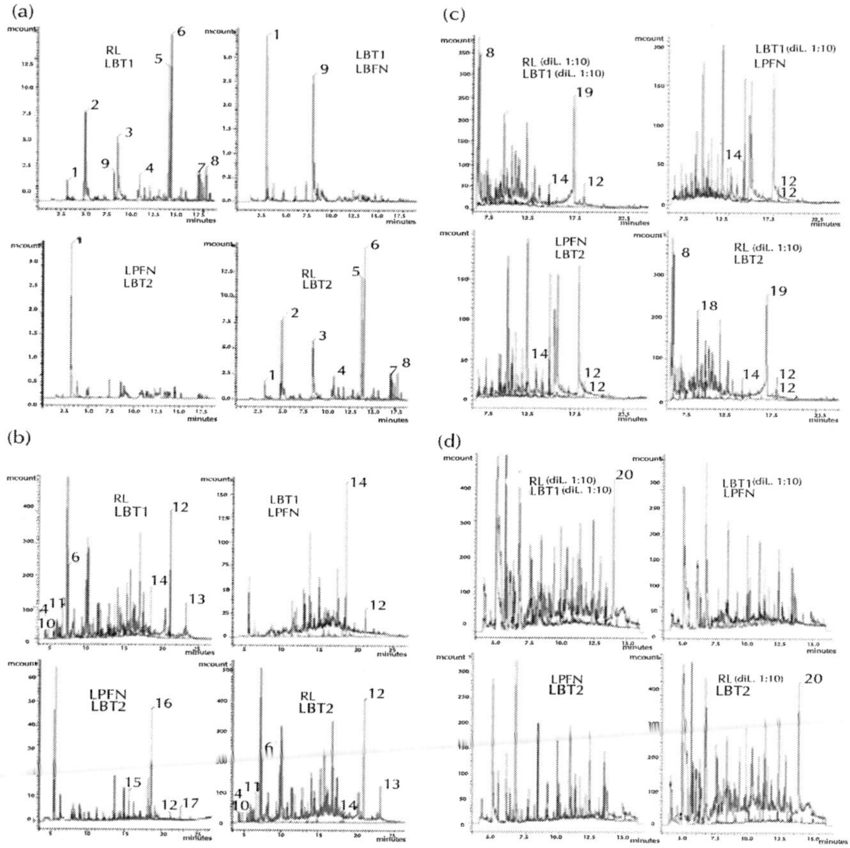

Figure 7: Non-target screening analysis of leachate samples, collected at different treatment points, by the four methods described in Section 2.7: (a) VOCs; (b) PAHs, PCBs and phthalates; (c) pesticides; (d) phenols. Identification of contaminants removed and formed during different treatment stages (correspondence between the compound and the respective peak number is displayed inTable 3) (RL – Raw Leachate; LBT1 – Leachate after 1st biological treatment; LPFN – Leachate after Photo-Fenton reaction and neutralization; LBT2 – Leachate after 2nd biological treatment (final effluent)).

Table 3: Contaminants identified in the different leachate samples collected along the treatment process, after target and non-screening analyses by the four methods described in Section 2.7: structural characterization, fitting probability and removal efficiency during the different treatment stages (RL – Raw Leachate; LBT1 – Leachate after 1st biological treatment; LPFN – Leachate after Photo-Fenton reaction and neutralization; LBT2 – Leachate after 2nd biological treatment (final effluent))

Peak	**Contaminant**	**Molecular-weight (g/mol)**	**CAS no.**	**Molecular formula**	**Structural formula**	**Fitting probability (%)**	**Area/concentration in different samples**	**Removal (+)/ formation (−) (%)**
Non-target screening (area)								
1	Ethanol, 1-methoxy-, acetate	118	4382-77-8	$C_5H_{10}O_3$	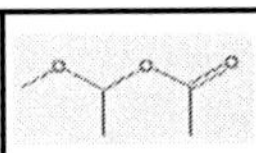	90	RL: 5.864e6	RL-LBT1: +77
						LBT1: 1.320e6	LBT1-LPFN: −925	
						LPFN: 1.353e7	LPFN-LBT2: +100	
						LBT2: n.d.[a]	RL-LBT2: +100	
2	Boronic acid, ethyl-, dimethyl ester	102	7318-82-3	$C_4H_{11}BO_2$	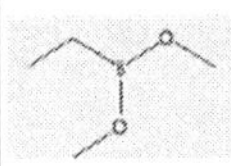	70	RL: 3.694e7	RL-LBT1: +99
						LBT1: 3.022e5	LBT1-LPFN: +100	
						LPFN: n.d.	LPFN-LBT2: n.a.[b]	
						LBT2: n.d.	RL-LBT2: +100	
3	Toluene	92	108-88-3	C_7H_8	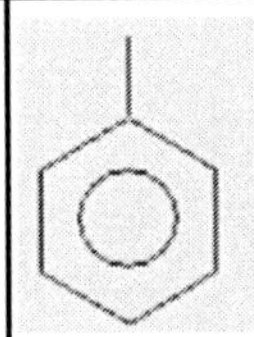	53	RL: 4.309e7	RL-LBT1: +92
						LBT1: 3.578e6	LBT1-LPFN: +43	
						LPFN: 2.046e6	LPFN-LBT2: +10	
						LBT2: 1.844e6	RL-LBT2: +96	

4	o-Xylene	106	95-47-6	C_8H_{10}	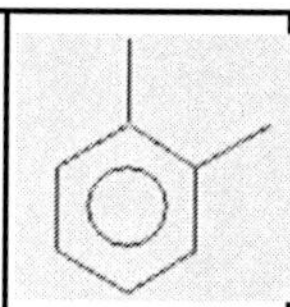	39	RL: 1.703e7	RL-LBT1: +94
						LBT1: 9.771e5	LBT1-LPFN: +100	
						LPFN: n.d.	LPFN-LBT2: n.a.	
						LBT2: n.d.	RL-LBT2: +100	
4′						RL: 1.661e8	RL-LBT1: +100	
						LBT1: n.d.	LBT1-LPFN: n.a.	
						LPFN: n.d.	LPFN-LBT2: n.a.	
						LBT2: n.d.	RL-LBT2: +100	
5	4-Isopropyltoluene	134	99-87-6	$C_{10}H_{14}$	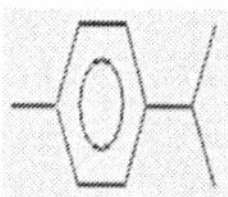	57	RL: 4.891e7	RL-LBT1: +98
						LBT1: 7.974e5	LBT1-LPFN: +100	
						LPFN: n.d.	LPFN-LBT2: n.a.	
						LBT2: n.d.	RL-LBT2: +100	
6	Eucalyptol	154	470-82-6	$C_{10}H_{18}O$	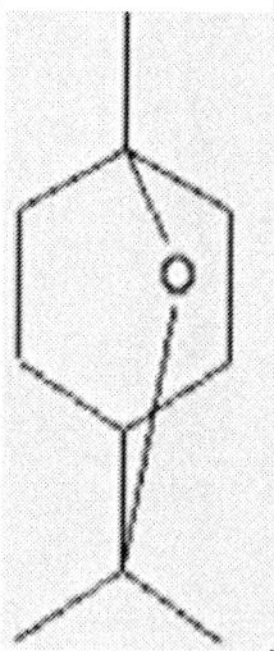	83	RL: 4.796e7	RL-LBT1: +99
						LBT1: 1.414e5	LBT1-LPFN: +100	
						LPFN: n.d.	LPFN-LBT2: n.a.	
						LBT2: n.d.	RL-LBT2: +100	
6′						RL: 4.275e8	RL-LBT1: +99	
						LBT1: 4.631e6	LBT1-LPFN: +54	
						LPFN: 2.118e6	LPFN-LBT2: +100	
						LBT2: n.d.	RL-LBT2: +100	
7	3-Cyclohexen-1-ol, 4-methyl-1-(1-methylethyl)-	154	562-74-3	$C_{10}H_{18}O$	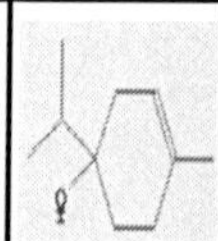	53	RL: 7.781e6	RL-LBT1: +96
						LBT1: 3.323e5	LBT1-LPFN: +23	
						LPFN: 2.546e5	LPFN-LBT2: +26	
						LBT2: 1.877e5	RL-LBT2: +98	

8	Cyclohexanone, 2-methyl-5-(1-methylethyl)-, trans	154	499-70-7	$C_{10}H_{18}O$	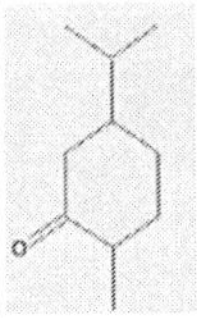	74		RL: 1.014e7	RL-LBT1: +97
							LBT1: 3.194e5	LBT1-LPFN: +26	
							LPFN: 2.371e5	LPFN-LBT2: +25	
							LBT2: 1.781e5	RL-LBT2: +98	
8′							RL: 4.710e9	RL-LBT1: +96	
							LBT1: 1.688e8	LBT1-LPFN: +76	
							LPFN: 4.060e7	LPFN-LBT2: +56	
							LBT2: 1.837e7	RL-LBT2: +99	
9	Disulfide, dimethyl	94	624-92-0	$C_2H_6S_2$	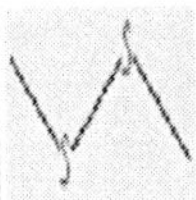	95		RL: 1.674e6	RL-LBT1: −753
						LBT1: 1.428e7		LBT1-LPFN: +100	
						LPFN: n.d.		LPFN-LBT2: n.a.	
						LBT2: n.d.		RL-LBT2: +100	
10	Ethylbenzene	106	100-41-4	C_8H_{10}	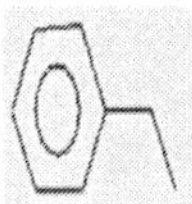	69		RL: 4.986e7	RL-LBT1: +100
						LBT1: n.d		LBT1-LPFN: n.a.	
						LPFN: n.d		LPFN-LBT2: n.a.	
						LBT2: n.d.		RL-LBT2: +100	
11	1,3,5-Trimethylbenzene	120	108-67-8	C_9H_{12}	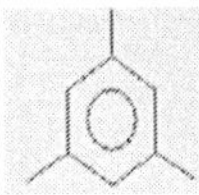	62		RL: 1.548e8	RL-LBT1: +78
						LBT1: 3.444e7		LBT1-LPFN: +97	
						LPFN: 1.150e6		LPFN-LBT2: +100	
						LBT2: n.d.		RL-LBT2: +100	

12	Bisphenol A or Phenol, 4,4'-(1-methylethylidene)bis-	228	80-05-7	$C_{15}H_{16}O_2$		86		RL: 1.097e9	RL-LBT1: +94
							LBT1: 6.615e7	LBT1-LPFN: +90	
							LPFN: 6.564e6	LPFN-LBT2: +39	
							LBT2: 4.023e6	RL-LBT2: +99	
12'							RL: 1.527e9	RL-LBT1: +90	
							LBT1: 1.581e8	LBT1-LPFN: +83	
							LPFN: 2.738e7	LPFN-LBT2: +17	
							LBT2: 2.267e7	RL-LBT2: +99	
13	1-Phenanthrene carboxylic acid, 1,2,3,4,4a,9,10,10a-octahydro-1,4a-dimethyl-7-(1-methylethyl)-, [1R-(1.alpha., 4a. beta., 10a.alpha.)]-	300	1740-19-8	$C_{20}H_{28}O_2$	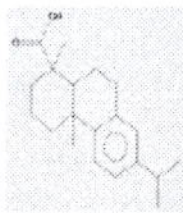	81		RL: 2.715e8	RL-LBT1: +100
							LBT1: n.d.	LBT1-LPFN: n.a.	
							LPFN: n.d.	LPFN-LBT2: n.a.	
							LBT2: n.d.	RL-LBT2: +100	
14	Phenol, 2,6-bis (1,1-dimethylethyl)-4-nitro-	251	728-40-5	$C_{14}H_{21}NO_3$	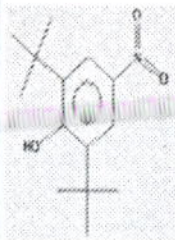	72		RL: n.d.	RL-LBT1: 100
							LBT2: 3.373e8	LBT1-LPFN: +100	
							LPFN: n.d.	LPFN-LBT2: −100	
							LBT2: 3.888e7	RL-LBT2: +100	
14'							RL: n.d.	RL-LBT1: −100	
							LBT1: 1.282e9	LBT1-LP-FN:+100	
							LPFN: n.d.	LPFN-LBT2: −100	
							LBT2: 1.143e8	RL-LBT2: +100	
15	4-Amino-7-diethylamino-chromen-2-one	232	107995-76-6	$C_{13}H_{16}N_2O_2$	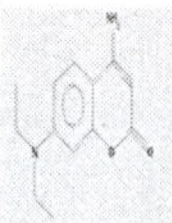	73		RL: 6.184e7	RL-LBT1:+53
							LBT1: 2.925e7	LBT1-LPFN: +38	
							LPFN: 1.967e7	LPFN-LBT2: +75	
							LBT2: 4.832e6	RL-LBT2: +92	
16	7,9-Di-tert-butyl-1-oxaspiro(4,5) deca-6,9-diene-2,8-dione	276	82304-66-3	$C_{17}H_{24}O_3$	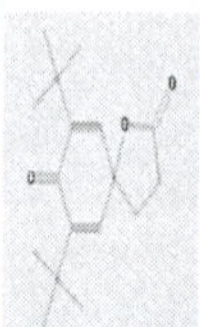	88		RL: n.d.	RL-LBT1: n.a.
							LBT1: n.d.	LBT1-LPFN: n.a.	
							LPFN: 7.097e7	LPFN-LBT2:+95	
							LBT2: 3.765e6	RL-LBT2: n.a.	

17	Benzyl butyl phtalate	312	85-68-7	$C_{19}H_{20}O_4$	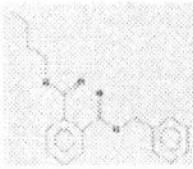	79 LBT1: n.d. LPFN: 9.528e6 LBT2: n.d.	RL: n.d. LBT1-LPFN: n.a. LPFN-LBT2:+100 RL-LBT2: n.a.	RL-LBT1: n.a.
18	Acenaphthene	154	83-32-9	$C_{12}H_{10}$	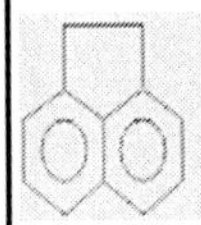	72 LBT1: 2.466e8 LPFN: 3.890e7 LBT2: 2.870e7	RL: 4.191e9 LBT1-LPFN: +84 LPFN-LBT2: +26 RL-LBT2: +99	RL-LBT1: +94
19	Cyclic octaatomic sulfur	256	10544-50-0	S_8	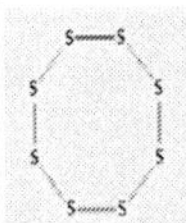	97 LBT1: n.d. LPFN: n.d. LBT2: n.d.	RL: 1.210e10 LBT1-LPFN: n.a. LPFN-LBT2: n.a. RL-LBT2: +100	RL-LBT1: +100
20	3,5-di-tert-Butyl-4-hydroxyphenyl-propionic acid	278	20170-32-5	$C_{17}H_{26}O_3$	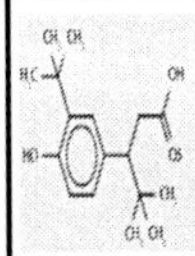	84 LBT1: n.d. LPFN: n.d. LBT2: n.d.	RL: 1.243e10 LBT1-LPFN: n.a. LPFN-LBT2: n.a. RL-LBT2: +100	RL-LBT1: +100
Target screening (Concentration in μg/L)								

Phenols	2-Nitrophenol	139	88-75-5	$C_6H_5NO_3$	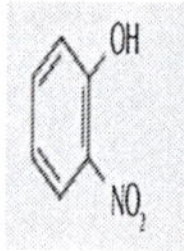	n.a. LBT1: 1.15 LPFN: 0.11 LBT2: 0.03	RL: 3.31 LBT1-LPFN: +90 LPFN-LBT2: +73 RL-LBT2: +99	RL-LBT1: +65
	2,4-Dimethylphenol	122	105-67-9	$C_8H_{10}O$	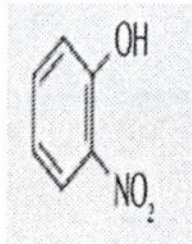	n.a. LBT1: 0.44 LPFN: n.d. LBT2: n.d.	RL: 3.21 LBT1-LPFN: +100 LPFN-LBT2: n.a.[b] RL-LBT2: +100	RL-LBT1: +86
	4-cloro-3-methylphenol	143	59-50-7	C_7H_7ClO	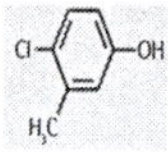	n.a. LBT1: 19.67 LPFN: 0.18 LBT2: 0.14	RL: 71.41 LBT1-LPFN: +99 LPFN-LBT2: +22 RL-LBT2: +99	RL-LBT1: +72
	p-tert-octylphenol	206	140-66-9	$C_{14}H_{22}O$	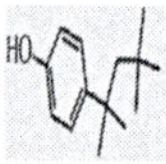	n.a. LBT1: 0.22 LPFN: n.d. LBT2: n.d.	RL: 1.51 LBT1-LPFN: +100 LPFN-LBT2: n.a. RL-LBT2: +100	RL-LBT1: +85
VOCs	Benzene	78	71-43-2	C_6H_6	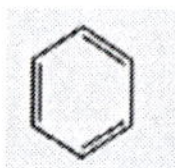	n.a. LBT1: n.d. LPFN: n.d. LBT2: n.d.	RL: 3.13 LBT1-LPFN: n.a. LPFN-LBT2: n.a. RL-LBT2: +100	RL-LBT1: +100
	1,2-Dichloroethane	99	107-06-2	$C_2H_4Cl_2$	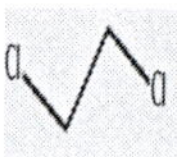	n.a. LBT1: n.d. LPFN: n.d. LBT2: n.d.	RL: 2.09 LBT1-LPFN: n.a. LPFN-LBT2: n.a. RL-LBT2: +100	RL-LBT1: +100
	1,2-Dichloropropane	113	78-87-5	$C_3H_6Cl_2$	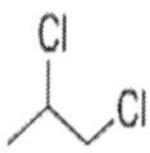	n.a. LBT1: n.d. LPFN: n.d. LBT2: n.d.	RL: 0.87 LBT1-LPFN: n.a. LPFN-LBT2: n.a. RL-LBT2: +100	RL-LBT1: +100

PAHs	Naphthalene	128	91-20-3	$C_{10}H_8$	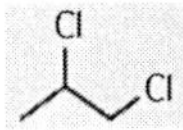	n.a. LBT1: n.d. LPFN: n.d. LBT2: n.d.	RL: 0.04 LBT1-LPFN: n.a. LPFN-LBT2: n.a. RL-LBT2: +100	RL-LBT1:+100
	Fluorene	166	86-73-7	$C_{13}H_{10}$	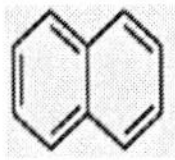	n.a. LBT1: n.d. LPFN: n.d. LBT2: n.d.	RL: 0.04 LBT1-LPFN: n.a. LPFN-LBT2: n.a. RL-LBT2: +100	RL-LBT1: +100
	Phenanthrene	178	85-01-8	$C_{14}H_{10}$	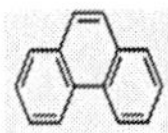	n.a. LBT1: 0.005 LPFN: n.d. LBT2: n.d.	RL: 0.06 LBT1-LPFN: +100 LPFN-LBT2: n.a. RL-LBT2: +100	RL-LBT1: +92
	Fluoranthene	202	206-44-0	$C_{16}H_{10}$	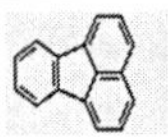	n.a. LBT1: n.d. LPFN: n.d. LBT2: n.d.	RL: 0.03 LBT1-LPFN: n.a. LPFN-LBT2: n.a. RL-LBT2: +100	RL-LBT1: +100
	Pyrene	202	129-00-0	$C_{16}H_{10}$	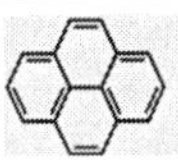	n.a. LBT1: n.d. LPFN: n.d. LBT2: n.d.	RL: 0.008 LBT1-LPFN: n.a. LPFN-LBT2: n.a. RL-LBT2: +100	RL-LBT1: +100
	Dibenzo[a,h]anthracene	278	53-70-3	$C_{22}H_{14}$	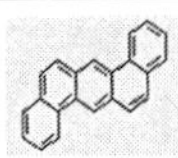	n.a. LBT1: n.d. LPFN: n.d. LBT2: n.d.	RL: 0.006 LBT1-LPFN: n.a. LPFN-LBT2: n.a. RL-LBT2: +100	RL-LBT1: +100

[a]n.d. – Not detected.

[b]n.a. – Not applicable.

Furthermore, the obtained results showed that many of the identified contaminants were originally relatively biodegradable, being significantly removed (>80%) after the first biotreatment process. However, the photo-Fenton reaction also played an important role for the leachate biodegradability increase (>70%), as stated by the >50% removal of several recalcitrant compounds (Table 3). One exception was spotted for compound corresponding to peak 1, identified as ethanol, 1-methoxy, acetate, which was produced in large amounts during the phototreatment stage (−975%), probably as a result of the degradation of other more complex aromatic compounds, as previously reported in literature. The 2nd biotreatment process efficiency was already quite low, which is in agreement with its aforementioned contribution to the elimination of the residual biodegradable organic carbon fraction.

CONCLUSIONS

A multistage treatment system consisting in a three sequential steps, activated sludge biological process/chemical oxidation system based on the photo-Fenton reaction/activated sludge biological process, showed to be a successful approach for the treatment of stabilized raw leachates from sanitary landfills, concerning the elimination of organic matter and nitrogen compounds.

Raw leachate presents a recalcitrant character, associated mainly to HS, and a high ammonium nitrogen content. Biological treatment with activated sludge, under aerobic and anoxic conditions, achieved 39% mineralization of the organic carbon and 95% reduction of the nitrogen content. The highest nitrification rate obtained was 8.2 mg NH_4^+–N/h/g of volatile suspended solids (VSS) (T = 26.9 °C; pH = 7.6), consuming 4.5 g $CaCO_3$ per liter of raw leachate or 1.2 mg $CaCO_3$ per mg NH_4^+–N. The maximum denitrification rate obtained was 5.8 mg (NO_2^-–N + NO_3^-–N)/h/g VSS (T = 26.4 °C; pH = 8.4), with a C/N consumption ratio of 2.4 mg CH_3OH per mg (NO_2^-–N + NO_3^-–N) (7.4 g/9.4 mL of commercial methanol per liter of leachate), with an overall alkalinity production of 4.3 g $CaCO_3$ per g (NO_2^-–N + NO_3^-–N) oxidized.

The phototreatment process led to the depletion of HS (>80%), of low-molecular-weight carboxylate anions (>70%) and other organic micropollutants, thus resulting in a total biodegradability increase to >70%, being possible to couple with a further biological treatment, achieving a final wastewater quality in agreement with discharge limits into receiving water bodies, with the exception of sulfate ions. Although humic acids and other organic compounds can be easily removed from the leachate by precipitation at pH ~3.0, leading to a high decrease of dissolved organic carbon and consequently decrease of the phototreatment time and reactants consumption, the elimination of the acid sludge can be a big concern.

ACKNOWLEDGMENTS

This work was partially supported by project PEst-C/EQB/LA0020/2011, financed by FEDER through COMPETE – Programa Operacional Factores de Competitividade and by FCT – Fundação para a Ciência e a Tecnologia. Financial support was also provided by EFACEC Engenharia and Sistemas, S.A. (Environment Unit). The authors further express their gratitude to IAREN – Water Institute of the Northern Region, as well as to FCT for T. Silva and M.A. Sousa doctorate scholarships (SFRH/BD/73510/2010 and SFRH/BD/44509/2008, respectively). V. Vilar acknowledges Ciência 2008 Program.

REFERENCES

1. Abbas, A.A., Jingsong, G., Ping, L.Z., Ya, P.Y., Al-Rekabi, W.S., 2009. Review on landfill leachate treatments. Journal of Applied Sciences Research 5 (5), 534e545.
2. Alleman, J.E., 1984. Elevated nitrite occurrence in biological wastewater treatment systems. Water Science and Technology 17, 409e419.

3. Bacardit, J., Oller, I., Maldonado, M.I., Chamarro, E., Malato, S., Esplugas, S., 2007. Simple models for the control of photoFenton by monitoring h2o2. Journal of Advanced Oxidation Technologies 10 (2), 219e228.
4. Baun, A., Ledin, A., Reitzel, L.A., Bjerg, P.L., Christensen, T.H., 2004. Xenobiotic organic compounds in leachates from ten Danish MSW landfillsechemical analysis and toxicity tests. Water Research 38 (18), 3845e3858.
5. Beceiro-Gonza´lez, E., Concha-Gran˜ a, E., Guimaraes, A., Gonc¸alves, C., Muniategui-Lorenzo, S., Alpendurada, M.F., 2007. Optimisation and validation of a solid-phase microextraction method for simultaneous determination of different types of pesticides in water by gas chromatographyemass spectrometry. Journal of Chromatography A 1141 (2), 165e173.
6. Canziani, R., Emondi, V., Garavaglia, M., Malpei, F., Pasinetti, E., Buttiglieri, G., 2006. Effect of oxygen concentration on biological nitrification and microbial kinetics in a cross-flow membrane bioreactor (mbr) and moving-bed biofilm reactor (mbbr) treating old landfill leachate. Journal of Membrane Science 286 (1e2), 202e212.
7. Chian, E.S.K., DeWalle, F.B., 1976. Sanitary landfill leachates and their treatment. Journal of the Environmental Engineering Division, 411e431.
8. Christensen, T.H., Kjeldsen, P., Bjerg, P.L., Jensen, D.L., Christensen, J.B., Baun, A., Albrechtsen, H.-J., Heron, G., 2001. Biogeochemistry of landfill leachate plumes. Applied Geochemistry 16 (7e8), 659e718.
9. Christensson, M., Lie, E., Welander, T., 1994. A comparison between ethanol and methanol as carbon sources for denitrification. Water Science and Technology 30, 83.
10. Clesceri, L.S., Greenberg, A.E., Eaton, A.D., 2005. Standard Methods for Examination of Water & Wastewater, 21st ed. American Public Health Association (APHA), American Water

Works Association (AWWA) & Water Environment Federation (WEF).

11. EMPA, 1992. OCDE Guideline for Testing of Chemicals, Adopted by the Council on 17th July 1992, Zahn-Wellens/ EMPA Test. Swiss Federal Laboratories for Materials testing and Research.
12. EPA, 1996. U.S. Environmental Protection Agency, Prevention Pesticides and Toxic Substances (7101). Fates; Transport and Transformation Test Guidelines Oppts 835.3200 Zahnwellens/ EMPA Test. EPA 712-C-96-084, Washington, DC.
13. Fan, H.-j., Shu, H.-Y., Yang, H.-S., Chen, W.-C., 2006. Characteristics of landfill leachates in central Taiwan. Science of the Total Environment 361 (1e3), 25e37.
14. Folin, O., Ciocalteau, V., 1927. On tyrosine and tryptophane determinations in proteins. Journal of Biological Chemistry 73, 627e650.
15. Ga´lvez, J.B., Rodrı´guez, S.M., 2003. Solar Detoxification. United Nations Educational, Scientific and Cultural Organization.
16. Gonc¸alves, C., Alpendurada, M.F., 2004. Solid-phase microextractionegas chromatographye(tandem) mass spectrometry as a tool for pesticide residue analysis in water samples at high sensitivity and selectivity with confirmation capabilities. Journal of Chromatography A 1026 (1e2), 239e250.
17. Gotvajn, A.Z., Tisler, T., Zagorc-Koncan, J., 2009. Comparison of different treatment strategies for industrial landfill leachate. Journal of Hazardous Materials 162 (2e3), 1446e1456.
18. Guimara˜es, A.D., Carvalho, J.J., Gonc¸alves, C., Alpendurada, M.D.F., 2008. Simultaneous analysis of 23 priority volatile compounds in water by solid-phase microextraction-gas chromatography-mass spectrometry and estimation of the method's uncertainty. International Journal of Environmental Analytical Chemistry 88 (3), 151e164.

19. Jensen, D.L., Ledin, A., Christensen, T.H., 1999. Speciation of heavy metals in landfill-leachate polluted groundwater. Water Research 33 (11), 2642e2650.
20. Jonsson, S., Ejlertsson, J., Svensson, B.H., 2003. Behaviour of mono- and diesters of o-phthalic acid in leachates released during digestion of municipal solid waste under landfill conditions. Advances in Environmental Research 7 (2), 429e440.
21. Kulikowska, D., Klimiuk, E., 2004. Removal of organics and nitrogen from municipal landfill leachate in two-stage SBR reactors. Polish Journal of Environmental Studies 13 (4), 389e396.
22. Lema, J.M., Me´ndez, R., Bla´zquez, R., 1988. Characteristics of landfill leachates and alternatives for their treatment: a review. Water, Air and Soil Pollution 40, 223e250.
23. Marttinen, S.K., Kettunen, R.H., Rintala, J.A., 2003. Occurrence and removal of organic pollutants in sewages and landfill leachates. Science of the Total Environment 301 (1e3), 1e12.
24. Metcalf and Eddy, 2005. Wastewater Engineering Treatment and Reuse, fourth ed. Metcalf & Eddy.
25. Modin, O., Fukushi, K., Yamamoto, K., 2007. Denitrification with methane as external carbon source. Water Research 41 (12), 2726e2738.
26. Nogueira, R.F.P., Oliveira, M.C., Paterlini, W.C., 2005. Simple and fast spectrophotometric determination of h2o2 in photoFenton reactions using metavanadate. Talanta 66, 86e91.
27. O¨ man, C.B., Junestedt, C., 2008. Chemical characterization of landfill leachates e 400 parameters and compounds. Waste Management 28 (10), 1876e1891.
28. Osako, M., Kim, Y.-J., Sakai, S.-I., 2004. Leaching of brominated flame retardants in leachate from landfills in Japan. Chemosphere 57 (10), 1571e1579.

29. Oturan, M.A., Pimentel, M., Oturan, N., Sire´s, I., 2008. Reaction sequence for the mineralization of the short-chain carboxylic acids usually formed upon cleavage of aromatics during electrochemical Fenton treatment. Electrochimica Acta 54 (2), 173e182.

30. Pignatello, J.J., Oliveros, E., MacKay, A., 2006. Advanced oxidation processes for organic contaminant destruction based on the Fenton reaction and related chemistry. Critical Reviews in Environmental Science and Technology 36, 1e84.

31. Prieto-Rodrı´guez, L., Oller, I., Zapata, A., Agu¨ era, A., Malato, S., 2011. Hydrogen peroxide automatic dosing based on dissolved oxygen concentration during solar photo-Fenton. Catalysis Today 161 (1), 247e254.

32. Primo, O., Rivero, M.J., Ortiz, I., 2008. Photo-Fenton process as an efficient alternative to the treatment of landfill leachates. Journal of Hazardous Materials 153 (1e2), 834e842.

33. Pulgarin, C., Invernizzi, M., Parra, S., Sarria, V., Polania, R., Pe´ringer, P., 1999. Strategy for the coupling of photochemical and biological flow reactors useful in mineralization of biorecalcitrant industrial pollutants. Catalysis Today 54 (2e3), 341e352.

34. Randall, C.W., Barnard, J.L., Stensel, H.D., 1992. In: Eckenfelder, W.W., Malina, J.F., Patterson, J.W. (Eds.), Design and Retrofit of Wastewater Treatment Plants for Biological Nutrient Removal. Technomic Publishing AG, Lancaster, Pensylvania.

35. Renou, S., Givaudan, J.G., Poulain, S., Dirassouyan, F., Moulin, P., 2008. Landfill leachate treatment: review and opportunity. Journal of Hazardous Materials 150 (3), 468e493.

36. Robinson, H.D., Maris, P.J., 1985. The treatment of leachates from domestic waste in landfill sites. Research Journal of the Water Pollution Control Federation 57, 30e38.

37. Rocha, E.M.R., Vilar, V.J.P., Fonseca, A., Saraiva, I., Boaventura, R.A.R., 2011. Landfill leachate treatment by solardriven aops. Solar Energy 85 (1), 46e56.

38. Ruiz, G., Jeison, D., Chamy, R., 2003. Nitrification with high nitrite accumulation for the treatment of wastewater with high ammonia concentration. Water Research 37 (6), 1371e1377.

39. Sawyer, C.N., McCarty, P.L., Parkin, G.F., 2003. Chemistry for Environmental Engineering and Science, fifth ed. McGraw-Hill, New York.

40. Silva, T.F.C.V., Silva, M.E.F., Cunha-Queda, A.C., Fonseca, A., Saraiva, I., Boaventura, R.A.R., Vilar, V.J.P., 2013a. Sanitary landfill leachate treatment using combined solar photoFenton and biological oxidation processes at pre-industrial scale. Chemical Engineering Journal 228, 850e866.

41. Silva, T.F.C.V., Fonseca, A., Saraiva, I., Vilar, V.J.P., Boaventura, R.A.R., 2013b. Biodegradability enhancement of a leachate after biological lagooning using a solar driven photo-Fenton reaction, and further combination with an activated sludge biological process, at pre-industrial scale. Water Research 47 (10), 3543e3557.

42. Spagni, A., Marsili-Libelli, S., 2009. Nitrogen removal via nitrite in a sequencing batch reactor treating sanitary landfill leachate. Bioresource Technology 100 (2), 609e614

43. Suthersan, S., Ganczarcczyk, J., 1986. Inhibition of nitrite oxidation during nitrification. Water Pollution Research Journal of Canada 21 (2), 257e266.

44. Thurman, E.M., Malcom, R.L., 1981. Preparative isolation of aquatic humic substances. Environmental Science & Technology 15, 463e466.

45. Vilar, V.J.P., Capelo, S.M.S., Silva, T.F.C.V., Boaventura, R.A.R., 2011a. Solar photo-Fenton as a pre-oxidation step for biological treatment of landfill leachate in a pilot plant with CPCS. Catalysis Today 161 (1), 228e234.

46. Vilar, V.J.P., Moreira, J.M.S., Fonseca, A., Saraiva, I., Boaventura, R.A.R., 2012a. Application of Fenton and solar photo-Fenton processes to the treatment of a sanitary landfill leachate in a pilot plant with CPCS. Journal of Advanced Oxidation Technologies 15 (1), 107e116.

47. Vilar, V.J.P., Rocha, E.M.R., Mota, F.S., Fonseca, A., Saraiva, I., Boaventura, R.A.R., 2011b. Treatment of a sanitary landfill leachate using combined solar photo-Fenton and biological immobilized biomass reactor at a pilot scale. Water Research 45 (8), 2647e2658.
48. Vilar, V.J.P., Silva, T.F.C.V., Santos, M.A.N., Fonseca, A., Saraiva, I., Boaventura, R.A.R., 2012b. Evaluation of solar photo-Fenton parameters on the pre-oxidation of leachates from a sanitary landfill. Solar Energy 86 (11), 3301e3315.
49. Villaverde, S., Ferna´ndez, M.T., Uruen˜a, M.A., Fdz-Polanco, F., 1997. Influence of substrate concentration on the growth and activity of a nitrifying biofilm in a submerged biofilter. Environmental Technology 18 (9), 921e928.
50. Wiszniowski, J., Robert, D., Surmacz-Gorska, J., Miksch, K., Malato, S., Weber, J.-V., 2004. Solar photocatalytic degradation of humic acids as a model of organic compounds of landfill leachate in pilot-plant experiments: influence of inorganic salts. Applied Catalysis B: Environmental 53 (2), 127e137.
51. Wiszniowski, J., Robert, D., Surmacz-Gorska, J., Miksch, K., Weber, J.V., 2006. Landfill leachate treatment methods: a review. Environmental Chemistry Letters 4, 51e61.
52. Wu, Y., Zhou, S., Qin, F., Peng, H., Lai, Y., Lin, Y., 2010. Removal of humic substances from landfill leachate by Fenton oxidation and coagulation. Process Safety and Environmental Protection 88 (4), 276e284.
53. Wu, Y., Zhou, S., Ye, X., Chen, D., Zheng, K., Qin, F., 2011. Transformation of pollutants in landfill leachate treated by a combined sequence batch reactor, coagulation, Fenton oxidation and biological aerated filter technology. Process Safety and Environmental Protection 89 (2), 112e120.
54. Zhang, H., Choi, H.J., Huang, C.-P., 2005. Optimization of Fenton process for the treatment of landfill leachate. Journal of Hazardous Materials 125 (1e3), 166e174.

55. Zhang, L., Li, A., Lu, Y., Yan, L., Zhong, S., Deng, C., 2009. Characterization and removal of dissolved organic matter (dom) from landfill leachate rejected by nanofiltration. Waste Management (Oxford) 29 (3), 1035e1040.

Chapter 4

Industrial Water Recycle/ Reuse

Jiří Jaromír Klemeš

Centre for Process Integration and Intensification – CPI[2], Research Institute of Chemical and Process Engineering, Faculty of Information Technology, University of Pannonia, Egyetem utca 10, 8200 Veszprém, Hungary

ABSTRACT

Water is used in most process industries for a wide range of applications. Industrial processes and systems using water are being subjected to increasingly stringent environmental regulations relating to the discharge of effluents. There is a growing demand for fresh water, which makes it precious in more and more countries very precious and in some parts of the world a crucial commodity. The changes and the pace of these changes have increased the need for improved water management and wastewater minimisation.

The adoption of water minimisation techniques can effectively reduce overall fresh water demand in water using processes and subsequently reduce the amount of effluent generated. This can result in reducing the cost incurred in the acquisition of fresh water and the cost of the treatment of effluent streams. This paper provides a brief overview of the recent techniques and methodologies, grouped in several topics: Water footprints and LCA; water/wastewater minimisation – including Water Pinch, Mathematical Programming techniques and combined water-energy minimisation. Case Studies demonstrating the significance of those techniques are briefly mentioned.

GRAPHICAL ABSTRACT

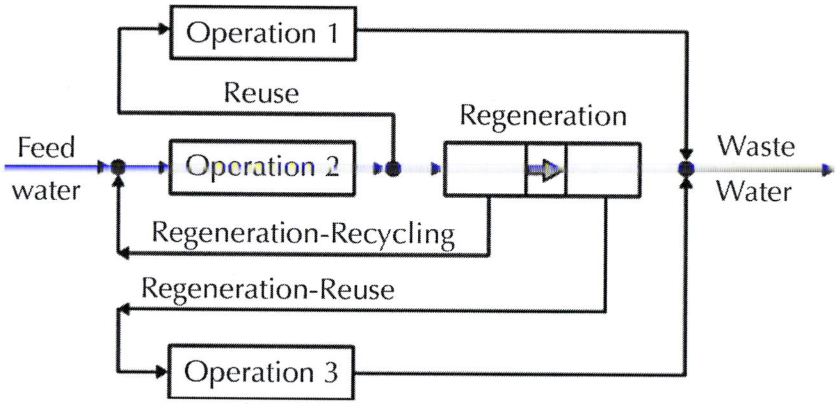

INTRODUCTION

There is a sustaining contribution of chemical engineering in this field. Important methodologies that have been applied in the minimisation of water use and effluent generation have included [1]:

- The minimisation of water consumption by the efficient management, best practices – best available techniques [2•] and control of process operations.
- The use of process integration techniques based on pinch analysis.
- The process optimisation based on mass and energy balances applying advanced optimisation strategies aiming at waste and effluent reduction represented, for example by multiobjective optimisation (MOO) based on mathematical programming.
- The integration of optimisation and production planning techniques in conjunction with real time plant measurements and control for product quality and minimisation of losses.
- The growing usages of enhanced intelligent support to operators by the application of knowledge-based decision-making procedures to select those options which best protect the environment.

Water use and wastewater discharge are both subject to national and international standards. Legislation in member countries of the European Union (EU) follows the directives of the European Commission (EC), which is the executive body of the EU. For IPPC Directive 96/61/EC (Directive 2008/1/EC codified version) see, for example [3].

In the USA is a number of federal statutes passed by Congress and signed into law by the President that are central to the Office of Water's mission. In addition, Presidential Executive Orders (EOs) play a central role in a number of Office of Water activities. EOs are legally binding orders that direct United States Environmental Protection Agency (EPA) and other federal agencies in their execution of Congressionally established laws and policies [4].

Similar legislations have got the other main world regions – China and various parts of Asia, Australia, Latin America and Africa. More information is available online and published elsewhere [5• and 6] and in the book edited by Atimay and Sikdar [7•].

WATER FOOTPRINT (WFP) AND LCA IN INDUSTRY

Hoekstra [8••] has defined the water footprint (WFP) as an indicator of direct and indirect water use. It has not been widely acknowledged how much water is needed for daily consumption of commons needs of the population. For example Chapagain and Hoekstra observe [9] that the water footprint of the Dutch tea consumption is 90 Mm^3/y and just the Dutch community (a country not with a large population) has a coffee-related water footprint of 2.6 billion m^3/y. Industrial water footprints have been very considerable.

The WFP is closely linked to the concept of virtual water [10] and represents the total volume of direct and indirect fresh water used, consumed, and/or polluted. A WFP consists of blue, green, and grey water footprints, which represent the consumption of surface and groundwater, the consumption of rainwater, and the volume of water required to dilute pollutants to water quality standards [11 and 12].

A review of Footprint Analysis Tools for Monitoring Impacts on Sustainability [13•] specified some important features of WFP and its relation to the variety of the other footprints. A WFP is an indicator for quantifying water usage for a particular product, for any well-defined group of consumers (e.g. an individual, industrial sire, city, province, state, or nation) or producers (e.g. a public organisation, industrial enterprise or even a sector. This issue has been further analysed elsewhere [14]. A WFP is measured in terms of water volumes consumed (evaporated or incorporated into the product) and polluted per unit of time or per functional unit [15]. The strength of the WFP concept is that it provides a broad perspective on the water management of the system, and allows for a deeper understanding of water usage. The WFP integrates water usage and pollution over the complete supply-chain [15]. On the contrary, the weaknesses of the WFP are in representation of just the quantity of water used. Without estimation of the related environmental impacts and with the lack of required data the estimation of grey

WFP is subjective [16]. Very limited uncertainty studies are so far available even though uncertainty can be significant [17•]. Amongst several other water footprints, the blue water footprint – BWF [18] deserves a special attention. It is a measure of the impact of water usage by individuals and societies on the world's freshwater resources. The BWF takes into account water consumption (water quantity), resource stress (the water stress index) and quality (the water impact index), see Grossman [19]. The BWF is expressed in volumetric units equivalent to water [20]. The water pollution footprint (WPF) represents the amounts of substances emitted into water [21].

To overcome some limitations of the footprint methodologies sustainable process index – SPI [22 and 23] and sustainable environmental performance indicator – SEPI [24• and 25] have been introduced beside the others.

WATER AND WASTE WATER MINIMISATION

Technologies and Techniques for Water Recycling

The development made during 1990s, and from that time even accelerating, in the area of water networks has been mainly related to the provision of targeting and design procedure based on graphical representation of water use and treatment. It has been effective for engineers to gain conceptual understanding and physical insights for water reuse. A range of optimisation frameworks had been proposed for solving large-size industrial problems in which a large number of operations are involved and with complex design interactions. They are parallel with the development of effective optimisation solvers [26•] to deal with multi-component problems with rigorous economic trade-offs. One key development of automated design was the integrated design framework which includes water-using systems and water-treating systems, and systematically evaluates simultaneously water reuse, regeneration recycling, regeneration reuse (Figure 1) and distributed treatment concept [27].

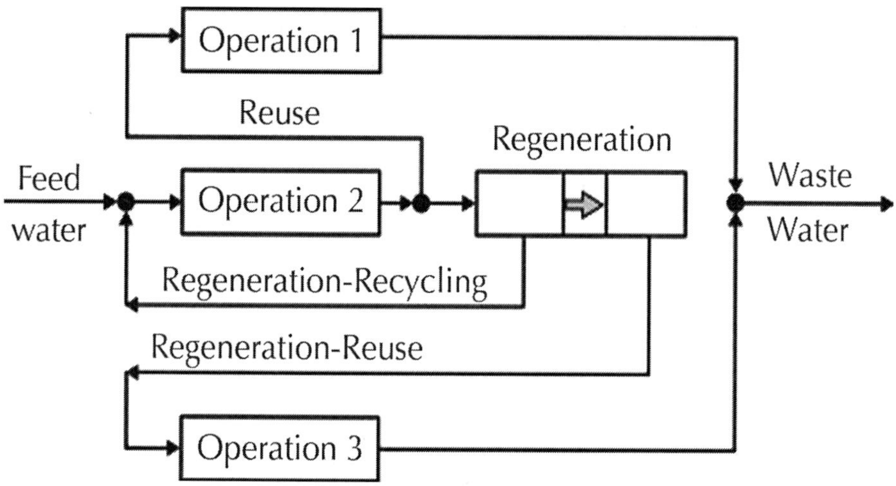

Figure 1: Potential water flows in Industry: water reuse, regeneration recycling, regeneration reuse.

Although significant advances have been made in the water network design by introducing optimisation techniques, operability of water network has not been fully addressed until recently. There exists operating uncertainty for the flowrate and concentration of water-using systems, and the practicality of network design obtained may not be ensured, or even the network is not feasible owing to fluctuation. This uncertain operating characteristic has been incorporated when water profile is defined and incorporated with the automated design framework [28], which enables realistic and practical design of water networks. Another issue added in the water network optimisation method recently is the prediction of pressure drop of the water network and its consideration for the design of pumping systems [29], with which economics of water networks are measured and optimised together with all the key elements in the costing, including freshwater, piping and pumping.

Water Pinch Analysis

The two main branches of Process Integration are energy and mass integration [30•, 31• and 32•]. Mass integration is a systematic methodology that provides a fundamental understanding of the global flow of mass within the process and then employs this understanding to identify performance targets and to optimise the allocation, separation, and generation of streams and species, developed initially by El-Halwagi and Manousiouthakis [33••] and Wang and Smith [34••]. In the context of wastewater minimisation, a mass integration problem involves transferring mass from rich process streams to lean process streams to achieve their target outlet concentration while simultaneously minimising waste generation and the consumption of utilities – including freshwater and external mass separating agents [35].

Pinch analysis was initiated for heat exchanger network synthesis [36•], and subsequently has been extended into other energy integration works [37•, 38 and 39••]. Water pinch analysis then emerged as a special case of mass integration (Figure 2). The original targeting technique was limited to the fixed load problem where water-using processes are modelled as mass transfer operations. Later works on water pinch analysis had been mainly dedicated to the fixed flowrate problems and extended in the further development [40].

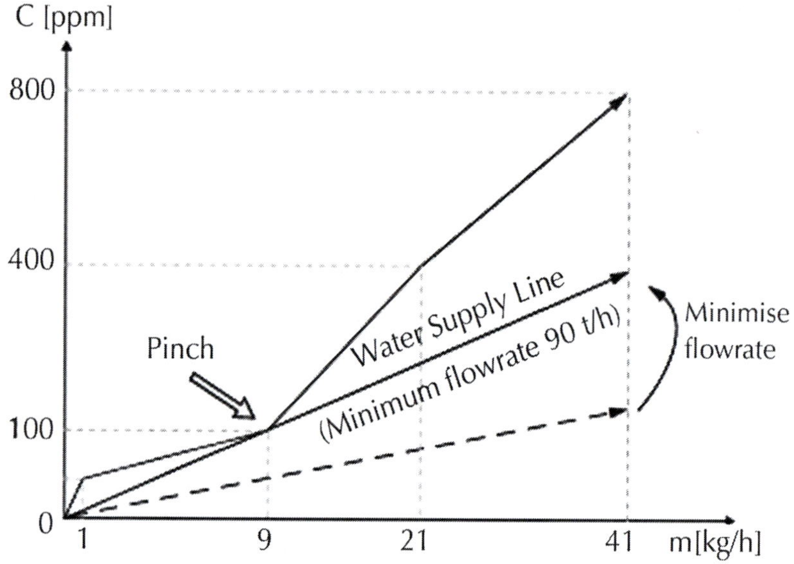

Figure 2: Limiting composite curve and the water pinch.

A typical pinch analysis study consists of two stages: setting minimum fresh water and wastewater flowrates (termed as targeting), followed by network design to achieve the targeted flowrates. The targeting enables to set a baseline target to determine how well a reuse/recycle system can actually perform. Knowing the targets ahead of design reduces problems with local optima. Once the minimum material targets are established, a water network can be designed using any network design tools. To set the flowrate targets for water reuse/recycle, various graphical and numerical targeting techniques may be employed. These include the Limiting Composite Curves [34••], Water Surplus Diagram [41], Material Recovery Pinch Diagram [42 and 43 •], Cascade Analysis Technique [44] and Source Composite Curve [45]. Various techniques can be used to design a water network that achieves the flowrate targets. A considerable contribution has been delivered by Deng and Feng [46•] and Deng *et al.* [47]. Several water pinch analysis works dedicated for batch processes have also been reported. These include works of Oliver *et al.* [48], Shoaib *et al.* [49•] and Nun *et al.* [50].

A minimum-water network methodology has been developed for water systems retrofit analysis using a systematic hierarchical approach for resilient process screening (SHARPS). This methodology considers a water cascade analysis for water targeting and a water management hierarchy to prioritise water system process changes [51 and 52•]. Iancu *et al.* [53] suggested waste water network retrofitting through optimal placement of regeneration unit.

Water Minimisation with Mathematical and Multiobjective Optimisation

Apart from water pinch analysis, mathematical optimisation techniques have also been utilised for water minimisation. Various optimisation approaches have been developed to complement water pinch analysis in dealing with more complex problems, for example multi-contaminant systems [54 and 55], complex operational constraints which include limiting the number of pipeline connections [56] and forbidden/compulsory matches between water-using processes [57].

Iancu *et al.* [58•] use a multi-objective function including the piping network cost, based upon the pipes' diameter, and the pumping cost, while the network length criterion is the sum of all the pipes through which contaminated water flows with a flow-rate higher than a threshold value. Iancu *et al.* [59] also proposed a pre-processing step for process design and optimisation of water networks, leading to equipartition of driving forces for mass transfer along of the associated system. The water network is abstracted as an oriented graph. Ahmetović and Grossmann [60•] contributed to the developed of strategies for the global optimisation of integrated process water networks. Bahiyah *et al.* [61] developed a systematic technique to target the minimum fresh water consumption and wastewater generation to achieve maximum water recovery (MWR) for systems involving multiple contaminants by generic linear programming (LP) model has been developed based on water network superstructure.

Combined Energy and Water – Simultaneous Minimisation

Water is often heated or cooled, according to operating temperature required by water-using operations. This requires certain external hot and cold utilities to water streams used in process industries. Water minimisation can be made with heat recovery of water use in a collective and holistic manner, which had been materialised with the graphical approach developed by Savulescu *et al.* [62•• and 63••] and Leewontanawit and Kim [64•]. This method allows targeting of minimum water and utility requirements together with heat recovery arrangement. The separate systems had been identified to simplify heat exchange matching between hot and cold streams. This is integrated with water networks based on maximum reuse complemented by the automated design method Leewontanawit and Kim [65]. This method is based on the optimisation for the combined superstructure of water networks and HENs (heat exchanger networks). The optimisation can be applied to the multi-component problems. A combination of both approaches graphical and MP was presented by Manan *et al.* [66•] for simultaneous minimisation of water and energy in process plants. Their technique consists of three steps: setting the minimum water and wastewater targets; design of minimum water utilisation network and finally the heat recovery network design. It is applicable to mass transfer based and non-mass transfer based water-using operations and it introduces a new graphical visualisation tool through a plot of temperature versus stream flowrate. Gololo*et al.* [67] presented a guided design of heating and cooling mains for lower water and energy consumption and increased efficiency. Burgara-Montero *et al.* [68] delivered optimal design of distributed treatment systems for the effluents discharged to rivers. Recent work of Boix *et al.* [69] used another two-step approach: an MILP is used to solve the water and energy allocation problem regarding several objectives and the second step improves the best results of the first phase with energy integration into the water network.

Short Overview of Several Case Studies

Just a few illustrative examples could be selected from hundreds available to demonstrate the variety of the reviewed methodologies implementations. Thevendiraraj *et al.* [70] successfully used the Water Pinch methodology to reduce the water wastewater generation by 22% for the maximum re-use analysis and 30% for the regeneration re-use consumption in citrus plant in Argentina. In Figure 3 is presented the network of the citrus plant and in Figure 4 is one of optimised variants. A case study on a brewery plant [57] and in the extended level [71] illustrated the use of a mathematical optimisation technique for water minimisation. In the brewery the volume ratio of water consumption to beer sold was reported as 6.04 L/L or 653 300 m^3/y. Joint integration of production and packaging sections results in considerable reduction of freshwater consumption by 37%. A notable reduction of waste water has been reported by Drobež *et al.* [72] for the production of biogas from organic and animal waste. Mateos-Espejel *et al.* [73] investigated the aspects of water and energy for the context of converting a Kraft mill into a forest biorefinery. Grabowski *et al.* [74] minimised energy and water use in a sugar manufacturing process based on cooling crystallisation of concentrated raw juice. Cortés *et al.* [75] used the heat integration followed by the mass integration for simultaneous energy and water minimisation applied to sugar process production. Saw *et al.* [76] applied an extended graphical targeting technique for direct reuse/recycle in concentration and property-based resource conservation networks including water and hydrogen networks. u ek *et al.* [77] studied combined energy, water and process technologies integration for the simultaneous production of ethanol and food from the entire corn plant. Khor *et al.* [78] used a nonlinear programming (NLP) model based on the superstructure and computational experiments using the GAMS/CONOPT3 modelling language platform for minimisation of water and waste water in a petroleum refinery. A recent work presented by Shukla *et al.* [79] dealt with the recycling of bleach plant effluent of an Indian paper mill using water cascade analysis technique. An interesting implementation of combined energy and water integration for an isolated site – an island was presented by Segurado *et al.* [80].

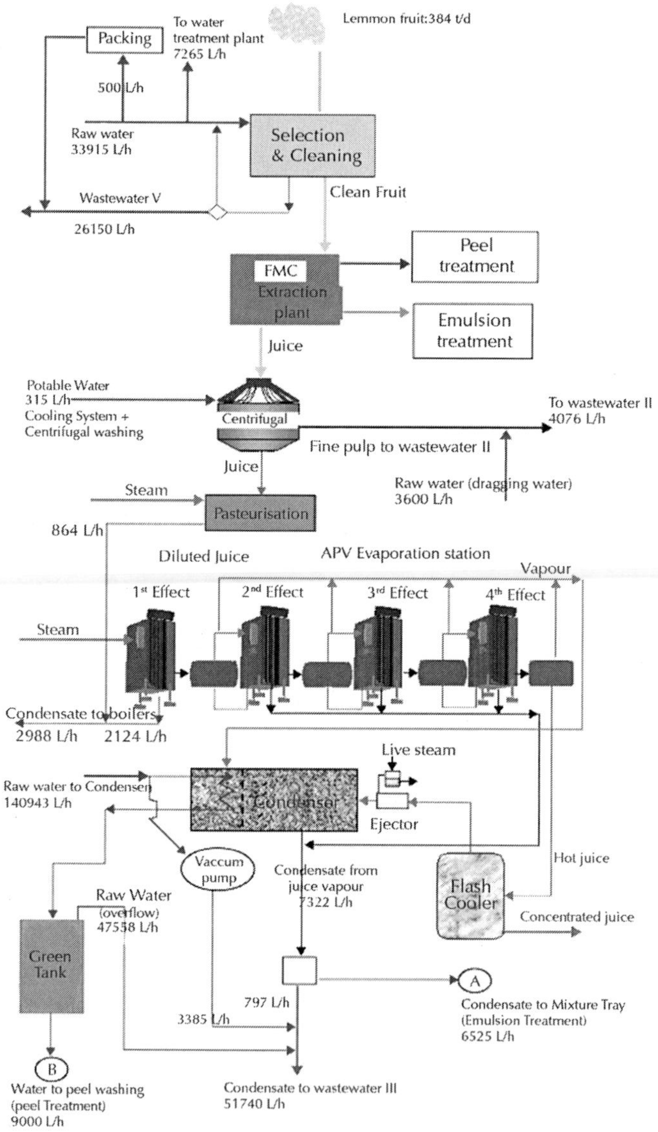

Figure 3: Water0020network of the citrus plant.

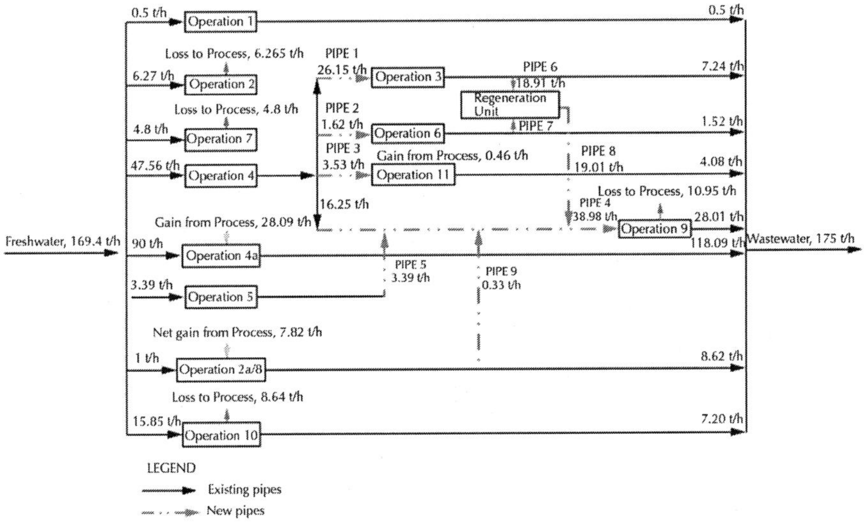

Figure 4: One water network optimised variants – result of water integration.

CONCLUSIONS

The development of the methodology for minimising water and waste water has been developing very fast and received considerable attention. It has been neither possible to provide more details nor the references owing to limited space available. Several books [30•, 31• and 32•] dealing partly or fully with topic as well as several comprehensive reviews [81•• and 82••] have been published. However owing to fast development in this field the reviews need to be updated regularly. Perhaps a good way how to follow the fast development of the research in water and wastewater minimisation is to attend relevant annual conferences. At least three of them should be mentioned: PRES – Process Integration, Modelling and Optimisation for Energy Saving and Pollution Reduction [83], ESCAPE – European Symposium on Computer Aided Process Engineering [84] and SDEWES – Conference on Sustainable Development of Energy, Water and Environment Systems [85].

REFERENCES

1. Klemesˇ J, Perry SJ: Process optimisation to minimise water use and wastage. In Waste Management and Co-product Recovery in Food Processing. Edited by Waldron K. Cambridge: Woodhead Publishing Limited; 2007:90-118.
2. 2.BAT/BREF, 2012, Common Waste Water and Waste Gas Treatment/Management Systems in the Chemical Sector, European Commission, Joint Research Centre, Institute for Prospective Technological Studies. Seville, Spain, www.eippcb.jrc.es/reference/cww.html, accessed 03.02.2012. A comprehensive material on Best Available Technique waste water treatment and minimisation in chemical industry.
3. EPA – Environmental Protection Agency, Wexford, Ireland, www.epa.ie/whatwedo/advice/bat/, accessed 13.01.2012.
4. US Environmental Protection Agency (EPA), www.water.epa.gov/ lawsregs/, accessed 21.03.2012.
5. Klemesˇ J, Smith R, Kim J-K (Eds): Handbook of Water and Energy Management in Food Processing. Cambridge, England: Woodhead Publishing Limited; 2008:1 84569 195 41029. An overview of water management and minimisation in food industry.
6. Klemesˇ J, Lam HL, Foo DCY: Water Recycling and recovery in food and drink industry. In Total Food, Sustainability of the AgriFoot Chain. Edited by Waldron K, Moates GK, Faulds CB. Cambridge: Royal Society of Chemistry; 2010:186-195.
7. Atimay AT, Sikdar SK: Security of Industrial Water Supply and Management, NATO Science for Peace and Security Series C: Environmental Security, 1–12. Dordrecht, the Netherlands: Springer; 2011 doi: 10.1007/978-94-007-1805-0_1. A collection of results dealing with the industrial water supply, management and optimisation.
8. Hoekstra AY: Water Neutral: Reducing and offsetting the impact of Water Footprints. The Value of Water Research Report Series 28. Delft, the Netherlands: UNESCO-IHE Institute for

Water Education; 2008. A key study and introduction of water footprints.

9. ChapagainAK, HoekstraAY: The waterfootprintof coffee andtea consumption in the Netherlands. Ecol Econ 2007, 64:109-111.
10. Hoekstra AY, Chapagain AK: Water footprints of nations: water use by people as a function of their consumption pattern. Water Resour Manag 2007, 21:35-48.
11. Mekonnen MM, Hoekstra AY: The green, blue and grey water footprint of farm animals and animal products. Value of Water Research Report Series No. 48. Delft, the Netherlands: UNESCOIHE; 2010.
12. Klemesˇ JJ, Varbanov P, Lam HL: Water footprint, water recycling and food industry supply chain. In Waste Management and Co-Product Recovery in Food Processing, vol 2. Edited by Waldron K. Cambridge: Woodhead Publishing Limited; 2009:978 1 84569 391 6134-168.
13. Cˇ ucˇ ek L, Klemesˇ JJ, Kravanja Z: A review of footprint analysis tools for monitoring impacts on sustainability. J Cleaner Prod 2012 doi: 10.1016/j.jclepro.2012.02.036. A review of various footprints and comparison of their impacts.
14. Cˇ ucˇ ek L, Klemesˇ JJ, Varbanov PS, Kravanja Z: Total footprintsbased multi-criteria optimisation of regional biomass energy supply chains. Energy 2012 doi: 10.1016/j.energy.2012.01.040.
15. Galli A, Wiedmann T, Ercin E, Knoblauch D, Ewing B, Giljum S: Integrating Ecological, Carbon and Water Footprint: Defining the "Footprint Family" and its Application in Tracking Human Pressure on the Planet. Surrey, UK: Technical Document; 2011.
16. Jeswani HK, Azapagic A: Water footprint: methodologies and a case study for assessing the impacts of water use. J Cleaner Prod 2011, 19:1288-1299.
17. Galli A, Wiedmann T, Ercin E, Knoblauch D, Ewing B, Giljum S: Integrating Ecological, Carbon and Water footprint into a

"Footprint Family" of indicators: definition and role in tracking human pressure on the planet. Ecol Indic 2012, 16:100-112. An interesting attempt to integrate footprints into a "footprint family".

18. Water Footprint Network. University of Twente, WWF-the global conservation organisation, UNESCO-IHE Institute for Water Education, the Water Neutral Foundation, the World Business Council for Sustainable Development, the International Finance Corporation (part of the World Bank Group) and the Netherlands Water Partnership, www.waterfootprint.org/?page=files/Glossary, accessed 13.02.2012.
19. Grossman BL: The Blue FootprintTM: Measuring the Critical Impact of Your Water Consumption; 2010, www.nanocleantechblog.com, accessed 24.08.11.
20. Auguste L: The Blue Footprint in Milwaukee, First-ever Water Impact Index and Carbon-Water-Economic Analysis, Veolia Water Americas; 2010, www.gmconline.org, accessed 27.08.11.
21. Sa´ nchez-Cho´ liz J, Duarte R: Water pollution in the Spanish economy: analysis of sensitivity to production and environmental constraints. Ecol Econ 2005, 53:325-338.
22. Sandholzer D, Narodoslawsky M: SPIonExcel – fast and easy calculation of the Sustainable Process Index via computer. Resour Conserv Recycl 2007, 50:130-142.
23. Kettl KH, Niemetz N, Sandor N, Eder M, Narodoslawsky M: Ecological impact of renewable resource-based energy technologies. J Fundam Renew Energy Appl 2011, 1 doi: 10.4303/jfrea/R101101 Article ID R101101.
24. De Benedetto L, Klemesˇ JJ: The environmental performance strategy map: an integrated LCA approach to support the decision making process. J Cleaner Prod 2009, 17:900-906. An introduction of an Environmental Performance Strategy Map combining various footprints.

25. De Benedetto L, Klemesˇ J: The environmental bill of materials and technology routing: an integrated LCA approach. Clean Technol Environ Policy 2010, 12:191-196.

26. Lam HL, Klemesˇ JJ, Kravanja Z, Varbanov P: Software tools overview: process integration, modelling and optimisation for energy saving and pollution reduction. Asia-Pacific J Chem Eng 2011, 6:696-712. Software tools overview covering a number of recent developments.

27. Gunaratnam M, Alva-Arga´ ez A, Kokossis A, Kim J-K, Smith R: Automated design of total water system design. Ind Eng Chem Res 2005, 44:588-599.

28. Hung SW, Kim J-K: Integrated water networks optimization under uncertainty. Chem Eng J 2011, 175:56-69.

29. Hung SW, Kim J-K: Optimization of water systems with the consideration of pressure drop and pumping. Ind Eng Chem Res 2012, 51:848-859.

30. Smith R: Chemical Process Design and Integration. UK: John Wiley & Sons, Ltd.; 2005. A comprehensive book covering the development of Energy and Water Integration until 2005.

31. El-Halwagi MM: Process Integration. Amsterdam, The Netherlands: Elsevier; 2006. A comprehensive book covering the development of Energy and Mass/ Water Integration until 2006.

32. Klemesˇ J, Friedler F, Bulatov I, Varbanov P: Sustainability in the Process industry – Integration and Optimization. New York: McGraw-Hill; 2010:. 362. A comprehensive book covering the development of Process Integration and optimisation for sustainability implementation in the industry up to 2010.

33. El-Halwagi MM, Manousiouthakis V: Synthesis of mass exchange networks. AIChE J 1989, 35:1233-1244. A key paper introducing mass/water integration methodology.

34. Wang Y, Smith R: Wastewater minimization. Chem Eng Sci 1994, 49:981-1006. A key paper introducing Water Integration and minimisation methodology.

35. Raskovic P: Water conservation in the chemical process industry – mass integration approach. In Security of Industrial Water Supply and Management, NATO Science for Peace and Security Series C: Environmental Security, 1–12. Edited by Atimay AT, Sikdar SK. Dordrecht, the Netherlands: Springer; 2011: 31-54.
36. Linnhoff B, Mason D, Wardle I: Understanding heat exchanger networks. Comput Chem Eng 1979, 3:295-302. One of key works on which Heat Integration is based.
37. Klemesˇ J, Dhole VR, Raissi IK, Perry SJ, Puigjaner L: Targeting and design methodology for reduction of fuel, power and CO2 on total sites. Appl Therm Eng 1997, 17:993-1003. An introduction of Total Site integration methodology.
38. Friedler F: Process integration, modelling and optimisation for energy saving and pollution reduction. Chem Eng Trans 2009, 18:1-26.
39. Friedler F: Process integration, modelling and optimisation for energy saving and pollution reduction. Appl Therm Eng 2010, 30:2270 2280. A comprehensive review of Process Integration and optimisation.
40. Dhole VR, Ramchandi N, Tainsh RA, Wasilewski M: Make your process water pay for itself. Chem Eng 1996, 103:100-103.
41. Hallale N: A new graphical targeting method for water minimisation. Adv Environ Res 2002, 6:377-390.
42. El-Halwagi MM, Gabriel F, Harell D: Rigorous graphical targeting for resource conservation via material recycle/reuse networks. Ind Eng Chem Res 2003, 42: 4319-4328.
43. Prakash R, Shenoy UV: Targeting and design of water networks for fixed flowrate and fixed contaminant load operations. Chem Eng Sci 2005, 60:255-268. One of key papers for fixed flowrate and fixed contaminants water networks.
44. Manan ZA, Foo CY, Tan YL: Targeting the minimum water flow rate using water cascade analysis technique. AIChE J 2004, 50:3169-3183.

45. Bandyopadhyay S, Ghanekar MD, Pillai HK: Process water management. Ind Eng Chem Res 2006, 45:5287-5297.

46. Deng C, Feng X: Targeting for conventional and propertybased water network with multiple resources. Ind Eng Chem Res 2011, 50:3722-3737. Introducing property based water-network and multiple resources.

47. Deng C, Feng X, Ng DKS, Foo DCY: Process-based graphical approach for simultaneous targeting and design of water network. AIChE J 2011, 57:3085-3104.

48. Oliver P, Rodrı´guez R, Udaquiola S: Water use optimization in batch process industries. Part 1: design of the water network. J Cleaner Prod 2008, 16:1275-1286.

49. Shoaib AM, Aly SM, Awad ME, Foo DCY, El-Halwagi MMA: Hierarchical approach for the synthesis of batch water network. Comput Chem Eng 2008, 32:530-539. One of key contributions to batch water network methodology.

50. Nun S, Foo DCY, Tan RR, Razon LF, Egashira R: Fuzzy automated targeting for trade-off analysis in batch water networks. Asia-Pacific J Chem Eng 2011, 6:537-551.

51. Wan Alwi SR, Manan ZA, Samingin MH, Misran N: Retrofit water systems the SHARPS way. Environ Manag 2006:20-27.

52. Wan Alwi SR, Manan ZA: SHARPS – a new cost-screening technique to attain cost-effective minimum water utilisation network. AIChE J 2006, 11:3981-3988. A successful development of retrofit methodology for water systems.

53. Iancu P, Plesu V, Lavric V: Waste water network retrofitting through optimal placement of regeneration unit. Chem Eng Trans 2009, 18:851-856.

54. Su W-N, Li Q-H, Liu Z-Y, Hui Pan Ch-H: A new design method for water-using network of multiple contaminants with single internal water main. J Cleaner Prod 2012 doi: 10.1016/j.jclepro.2012.01.041.

55. Fan X-Y, Li Y-P, Liu Z-Y, Pan Ch-H: A new design method for water-using networks of multiple contaminants with

the concentration potential concepts. Chem Eng Sci 2012, 73:345-353.

56. Hul S, Tan RR, Auresenia J, Fuchino T, Foo DCY: Water network synthesis using mutation-enhanced particle swarm optimization. Process Saf Environ Prot 2007, 85:507-514.

57. Tokos H, Novak Pintarič Z: Synthesis of batch water network for a brewery plant. J Cleaner Prod 2009, 17:1465-1479.

58. Iancu P, Plesu V, Lavric V: Cost versus network length criteria in water network optimal design. Comput Aided Chem Eng 2006, 21:1821-1826. Introduction of cost of piping into the analysis making the studies more realistic.

59. Iancu P, Plesu V, Lavric V: Optimisation of water networks using ranking as pre-processing step to get driving force equipartition. Chem Eng Trans 2010, 21:289-294.

60. Ahmetović E, Grossmann IE: Strategies for the global optimization of integrated process water networks. Comput Aided Chem Eng 2010, 28:901-906. An attempt to tackle problem of global optimisation by nonlinear programming.

61. Bahiyah Z, Wan Alwi SR, Hashim H, Manan ZA: Optimal design of water networks involving multiple contaminants for global water operations, resource conservation. Asia Pacific Chem Eng J (APJ) 2011, 6:771-777.

62. Savulescu L, Kim J-K, Smith R.: Studies on simultaneous energy and water minimisation, Part I: systems with no water re-use. Chem Eng Sci 2005, 60:3279-3290. A key work in simultaneous energy and water integrations – no water reuse systems.

63. Savulescu L, Kim J-K, Smith R: Studies on simultaneous energy and water minimisation, Part II: systems with maximum re-use of water. Chem Eng Sci 2005, 60:3291-3308. A key paper introducing mass/water integration methodology Maximum water re-use systems.

64. Leewontanawit B, Kim J-K: Improving energy recovery for water minimisation. Energy 2009, 34:880-893. A successful

attempt for energy recovery with simultaneous water minimisation.

65. Leewontanawit B, Kim J-K: Synthesis and optimisation of heatintegrated multiple-contaminant water systems. Chem Eng Process 2008, 47:670-694.
66. Manan ZA, Tea SW, Wan Alwi SR: A new technique for simultaneous water and energy minimisation in process plant. Chem Eng Res Des 2009, 87:1509-1519. An extension of simultaneous water and energy minimisation in industry.
67. Gololo V, Majozi T, Zhelev T, Semkov K: Guided design of heating and cooling mains for lower water and energy consumption and increased efficiency. Chem Eng Trans 2011, 25:755-760.
68. Burgara-Montero O, Ponce-Ortega JM, Serna-Gonza´ lez M, ElHalwagi MM: Optimal design of distributed treatment systems for the effluents discharged to the rivers. Clean Technol Environ Policy 2012 doi: 10.1007/s10098-012-0469-2.
69. Boix M, Pibouleau L, Montastruc L, Azzaro-Pantel C, Domenech S: Minimizing water and energy consumptions in water and heat exchange networks. Appl Therm Eng 2012, 36:442-455.
70. Thevendiraraj S, Klemesˇ J, Paz D, Aso G, Cardenas J: Water and wastewater minimisation study of a citrus plant. Resour Conserv Recycl 2003, 37:227-250.
71. Tokos H, Novak-Pintaric Z, Yang Y, Kravanja Z: Multilevel strategies for the retrofit of large-scale industrial water system: a brewery case study. AIChE J 2012, 58:884-898.
72. Drobezˇ R, Novak-Pintaricˇ Z, Pahor B, Kravanja Z: Simultaneous heat integration and the synthesis of biogas process from animal waste. Asia-Pacific J Chem Eng 2010, 6:734-749.
73. Mateos-Espejel E, Alva-Argaez A, Savulescu L, Paris J: From Kraft mills to forest biorefinery: an energy and water perspective, Part I: methodology. Cellulose Chem Technol 2010, 44:15-19.

74. Grabowski M, Klemesˇ J, Urbaniec K, Vaccari G, Wernik J: Energy and water use in a sugar manufacturing process based on cooling crystallisation of concentrated raw juice. Zuckerindustrie 2002, 127:604-609.
75. Corte´ s MG, Verelst H, Pedraja RE, Sua´rez EG: Simultaneous energy and water minimization applied to sugar process production. Chem Eng Trans 2011, 25:177-182.
76. SawSY, Lee L, LimMH, Foo DCY, ChewIML, TanRR,Klemesˇ JJ:An extended graphicaltargeting technique for directreuse/ recycle in concentration and property-based resource conservation networks. Clean Technol Environ Policy 2011, 13:347-357.
77. Cˇ ucˇ ek L, Martin M, Grossmann I, Kravanja Z: Energy, water and process technologies integration for the simultaneous production of ethanol and food from the entire corn plant. Comput Chem Eng 2011, 35:1547-1557.
78. Khor CS, Shah N, Mahadzir S, Elkamel A: Optimisation of petroleum refinery water network systems retrofit incorporating reuse, regeneration and recycle strategies. Can J Chem Eng 2012, 90:137-143.
79. Shukla SK, Kumar V, Yeom IT, Bansal MC: Recycling of bleach plant effluent of an Indian paper mill using water cascade analysis technique. Clean Technol Environ Policy 2011 doi: 10.1007/s10098-011-0432-7.
80. Segurado R, Alves LM, Duic´ N, Krajacˇ ic´ G: Integrated Energy and Water Planning on an Arid Island, Case of S. Vicente, Cape Verde. Jordan J Mech Ind Eng 2010, 4:1–8, www.jjmie.hu.edu.jo/ files/v4n1/1.pdf, accessed 29.02.2012.
81. Jez˙owski J: Review of water network design methods with literature annotations. Ind Eng Chem Res 2010, 49:4475-4516. A comprehensive review of water network design.
82. Foo DCY: State-of-the-art review of pinch analysis techniques for Water network synthesis. Ind Eng Chem Res 2009, 48:5125-5159. A comprehensive state-of-the art review of Process Integration (Pinch) methodology.

83. PRES – Process Integration for Energy Saving and pollution Reduction, www.conferencepres.com, accessed 03.02.2012.
84. ESCAPE, European Symposium on Computer Aided Process Engineering, www.cape-wp.eu/page.php?30, accessed 29.02.2012.
85. SDEWES, www.ohrid2012.sdewes.org/, accessed 01.03.2012.

Chapter 5

Catalytic Hydrothermal Treatment of Pharmaceutical Wastewater Using Sub- and Supercritical Water Reactions

Shirin Falamarzian[a], Omid Tavakoli[a], Reza Zarghami[a], and Mohammad Ali Faramarzi[b]

[a]School of Chemical Engineering, College of Engineering, University of Tehran, Tehran 14176, Iran

[b]Department of Pharmaceutical Biotechnology, Faculty of Pharmacy & Biotechnology Research Center, Tehran University of Medical Sciences, Tehran 1417614411, Iran

ABSTRACT

Application of subcritical and supercritical water technology for destruction of pharmaceutical compounds (carbamazepine, metoprolol and sulfamethaxazole) was investigated. The experiments were conducted inside batch reactor at a temperature ranging from 473 to 773 K and with different residence times of 5 to 50 min. The results show that carbamazepine, metoprolol and sulfamethaxazole are destructed by 90.27%, 99.99% and 98.84% after a 20 min exposure to 623 K, 673 K and 573 K, respectively. In comparison with the conventional methods of pharmaceutical waste treatment, the current technology provides a higher destruction efficiency (approximately 90–100%) which is achievable in shorter durations. NaOH and $CuSO_4{\cdot}5H_2O$ were also applied as catalysts in the temperature range of 473 K to 723 K. Comparing these catalysts, $CuSO_4{\cdot}5H_2O$ demonstrates a higher destruction efficiency, especially at lower temperatures. Based on the proposed pathway, the products of destructioncan be classified as environmentally-friendly compounds. The results show that this technology can be used as a green alternative for efficient removal of pharmaceutical compounds from wastewater streams.

GRAPHICAL ABSTRACT

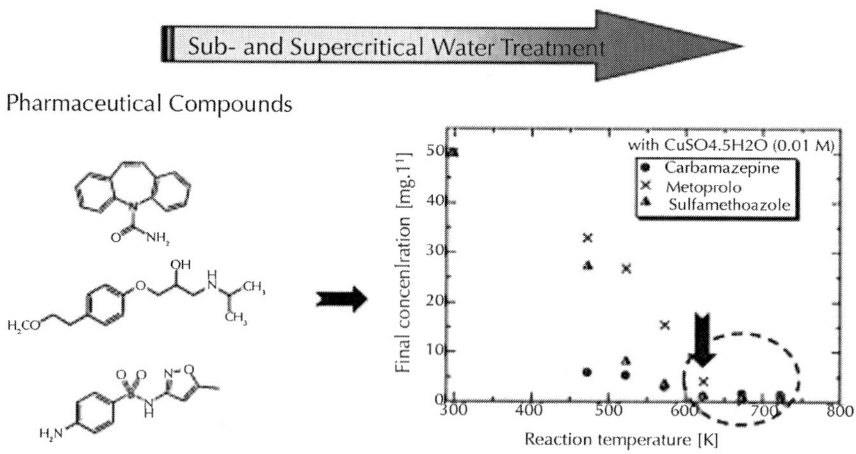

INTRODUCTION

The presence of pharmaceuticals in wastewater streams, surface waters, ground waters and related soil environments can have undesirable effects and has therefore attracted a great deal of attention in recent years. As a result,researchers have focused on devising new approaches for wastewater treatment [1], [2],[3], [4] and [5]. The presence of these organic contaminants in the environment stems from inadequate pharmaceutical compound treatment in wastewater treatment plants (WWTPs) and may lead to chronic changes and ecotoxicity [3].

Many studies have focused on the ineffectiveness of conventional contaminated water treatment facilities. In one study, some -blockers like metoprolol, propranolol, sotalol and psychiatric drugs such as carbamazepine and hormones were shown to be present in the effluent of a water treatment plant in Spain [3]. Behara et al. studied 5 wastewater treatment plants in South Korea using conventional biological treatment methods and found inefficient removal of 20 pharmaceuticals [2]. Another research on 57 pharmaceutical compounds of 4 wastewater treatment plants in Taiwan showed low removal efficiencies for most drugs, including atenolol, sulfamethoxazole, carbamazepine and metoprolol [4]. There are many other studies elucidating the low removal efficiencies of pharmaceutical compounds such as chloramphenicol, metoprolol, carbamazepine, sulfamethazine, sulfamethoxazole and diazepam [6], [7] and [8]. The continuous discharge of antibiotics from WWTP and their adverse effects on human health by promoting microbial drug resistance have raised concern about their adverse effects in the environment [9].

An alternative wastewater treatment technology that has gained much attention in recent years is sub-critical and supercritical water technology which is based on the unique behavior of water in its near-critical, critical (T_c = 647.1 K, P_c = 22.1 MPa, c = 0.322 g/cm^3) and supercritical regions [10]. Raising the self-association of water (kw) as a result of increasing temperature has severe effects on hydrolysis and acid-base equilibrium; therefore,water can act as an acid or base catalyst precursor because of its high content of H_3O^+ and OH^- ions [10] and [11]. Increasing temperature also brings about increasing viscosity ( ) at gas-like densities and decreasing at liquid-like densities which result in high fluidity, high molecular mobility and, subsequently, high thermal conductivity of water [10]. Dielectric constant (ε) of water, which shows solvent behavior and ionic dissociation of salts in water, decreases by increasing temperature and decreasing density (a non-polar state of water), causing an increase in the solubility of hydrophobic organic compounds (HOC) at such conditions [10], [12], [13], [14] and [15]. Briefly, supercritical water is an excellent solvent for homogeneous media without phase boundaries and also provides fast and complete reactions [16] and [17]. Treatment of some materials such as methane [18], [19] and [20], methanol [21], [22], [23], [24] and [25], ethanol [26] and [27], propane [28], nitrogen [29] and phenol [30], [31], [32] and [33] has been conducted in subcritical and supercritical water. Furthermore, efficient removal of some compounds including benzene [34], biphenyls [35], [36], [37], [38] and [39], amines [40] and [41] and pyridine [42] and [43] have also been studied in subcritical and supercritical water. Finally, this process can be considered as a green and environmental-friendly technology since it emits no harmful materials to the environment [17], [35], [39], [44] and [45].

In this study, sub- and supercritical water technology was applied to water contaminated with three pharmaceutical compounds; carbamazepine, metoprolol and sulfamethoxazole, and the effect of changing temperature and time on destruction of these compounds was examined in a batch reactor. Moreover, the effects of potential catalysts, NaOH and $CuSO_4{\cdot}5H_2O$ in destruction efficiency were studied.

MATERIALS AND METHODS

Chemicals

The 99.8% purified carbamazepine and metoprolol used in this experimental study was obtained from Loghman Co., Pursina Co. and Pars Daroo Co. (Tehran, Iran). Sulfamethoxazolewas was purchased from RoozDaroo Co. (Tehran, Iran). Fig. 1 shows the chemical structure of the above-mentioned compounds. H_2O used throughout the study was distilled and deionized by a laboratory apparatus. Sodium hydroxide (97% pure), copper (II) sulfate pentahydrate ($CuSO_4{\cdot}5H_2O$) (99% pure) were obtained from Merck (New Jersey, USA).

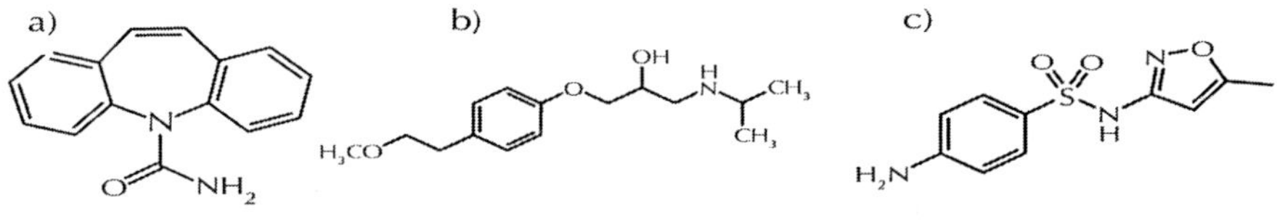

Figure 1: Chemical structure of (a) carbamazepine, (b) metoprolol and (c) sulfamethoxazole.

Experimental Apparatus

A tubular batch reactor made of stainless steel 316 SUS (volume 42 mL; length 29.48 cm; i.d. 2.13 cm) was built for this study. Swedgelok caps were used to reach the desired pressure and temperature based on thermodynamic calculations. The aqueous solution containing selected pharmaceutical compounds (initial concentration of 50 mg/L) with an amount of 5.0 to 30.0 mL (based on the pressure and temperature) were charged into the reactor. The reactor was then heated using an electrical furnace. The reactions were conducted in the temperature range of 498 to 773 K and pressure range of 1.5 to 30.0 MPa. After a certain reaction time, the reaction was terminated by immersing the reactor inside ice and water mixture until reaching room temperature.

Sample Preparation

After hydrothermal destruction of pharmaceutical compounds with and without catalyst (NaOH 1.0 M, $CuSO_4 \cdot 5H_2O$ 0.01 M), the liquid content was removed, centrifuged and analyzed. By analyzing different temperatures,the optimum temperature for each compound was determined and consequently, the furnace temperature was set on the optimum temperature for each compound and the experiments were done at various residence times of 5, 10, 20, 35 and 50 min.

Analysis

The HPLC apparatus (Knauer, Berlin, Germany) consisted of a Smartline HPLC Pump 1000 and was equipped with a PDA Detector 2800 (set at 254 nm), and a Degasser 5000 was used as the liquid content. Each sample (20 μL) was injected using a Smartline Autosampler 3950 with a sample loop of 100 μL. The data were acquired and processed by means of ChromGate software (version 3.3.1). Chromatographic separation was performed on a Lichrospher 100 RP & EC C18 reverse-phase column (C18, 25 × 0.46 cm i.d., 5 μM particle size) from Teknokroma (Barcelona, Spain) using an isocratic mobile phase of acetonitrile/ammonium acetate 0.1 M (for carbamazepine), methanol/water and 0.2% acetic acid (for metoprolol) and acetonitrile/50 mM phosphate buffer (K_2HPO_4, KH_2PO_4) for sulfamethoxazole. The retention times of carbamazepine, metoprolol and sulfamethoxazole were 6.0, 6.5, and 5.0 min, respectively.

RESULTS AND DISCUSSION

Effect of Temperature on Destruction of Pharmaceutical Compounds in the Absence of Catalyst

The effect of temperature (in the range of 523 to 773 K) on final concentration of selected pharmaceutical compounds at residence time of 30 min is shown in Fig. 2. As shown in the figure, the final concentration of the compounds decreased by enhancing the temperature with destruction efficiency of 90.5% (623 K), 100% (723 K) and 100% (573 K) for carbamazepine, metoprolol and sulfamethoxazole, respectively. The maximum efficiency for carbamazepine and sulfamethoxazole was obtained under sub-critical condition, whereas the maximum destruction of metoprolol occurred at supercritical temperature. The reason for reaching such efficiency at temperatures near critical point of water could be explained by enhancing the organic compounds solubility in water due to the decreasing dielectric constant of water which was already boosted by hydrolysis reaction power based on maximum ion product of water (kW). The figure also provides the pressure changes under reaction condition were taken in experiments.

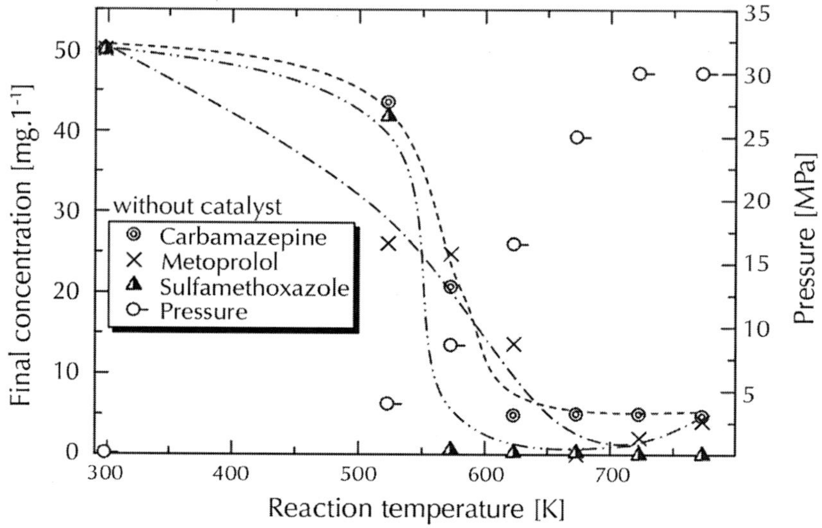

Figure 2: Effect of temperature and pressure on the final concentration of pharmaceutical compounds at 30 min and with no catalyst.

Effect of Residence Time on Destruction of Pharmaceutical Compounds in the Absence Catalyst

Fig. 3 illustrates the effect of different residence times of 5 to 50 min at fixed optimum temperature of 623 K.The conversions of pharmaceutical compounds increased by enhancing the residence time of reactions from 5 min to 20 min. The optimum residence time for all compounds was found to be 20 min and conversion of carbamazepine, metoprolol and sulfamethoxazole reached 91%, 96% and 96.4%, respectively. As shown, decreasing intensity in the range of 5 to 20 min for sulfamethoxazole was higher than that of the other two compounds due to its resistance to destruction at lower temperatures which gets back to its structure. However, comparing these results and the studies done by Jelic et al. (biological treatment of carbamazepine and metoprolol with 11% and 0% efficiency over a period of 16 days) [1], Lin et al. (activated sludge treatment of carbamazepine (40%), metoprolol (66%) and sulfamethoxazole (26%) in 12 h) [5] and Yu et al. (biodegradation of sulfamethoxazole (59%) over a period of 14 days) [46], showed that sub- and supercritical water treatment has high removal efficiency at low residence times.

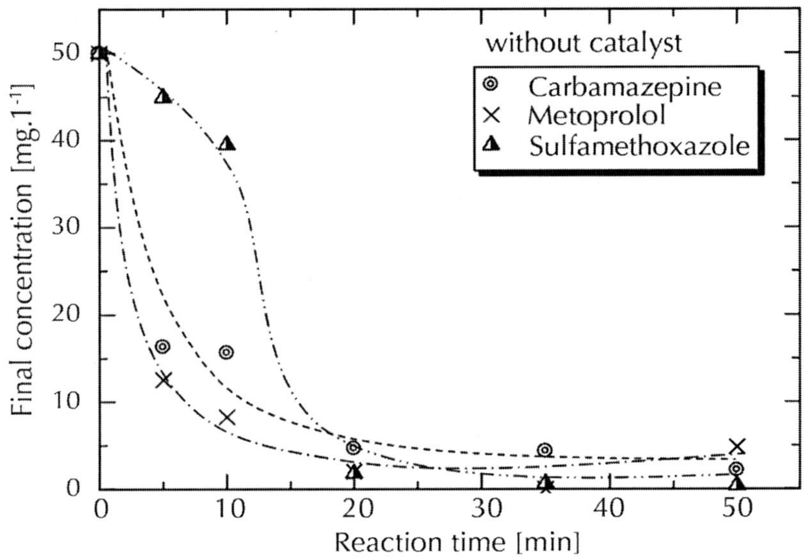

Figure 3: Effect of residence time on the final concentration of pharmaceutical compounds at 623 K (350 °C) and with no catalyst.

Effect of Catalyst on Destruction of Pharmaceutical Compounds

Catalytic Effect of NaOH and $CuSO_4$•$5H_2O$ on Carbamazepine Destruction

The catalytic effects of NaOH (1.0 M) and $CuSO_4 \cdot 5H_2O$ (0.01 M) on destruction of carbamazepine as a function of reaction temperature for a reaction time of 30 min is shown in Fig. 4 and the results of these experiments was compared to that of catalyst-free conditions. It is obvious that the presence of catalyst enhances the destruction of carbamazepine. Comparing the two catalysts, $CuSO_4$ illustrated higher destruction efficiency at lower temperature (88.5% at 473 K for 30 min compared to 15.7% with NaOH) while at higher temperatures above 500 K, both catalysts reached similar efficiency (maximum of 98% at 623 K). The obtained results show that using catalysts makes the destruction phenomenon occur at milder hydrothermal condition with less energy consumption. However, since NaOH caused severe corrosion in the reactor,to avoid damage to the reactor, this catalyst was not considered for carbamazepine and the other two pharmaceutical compounds. Based on the above-mentioned facts and from an economical viewpoint, using $CuSO_4{\cdot}5H_2O$ as a catalyst could enhance the destruction efficiency at lower temperature (94.5% at 573 K and 98% at 623 K).

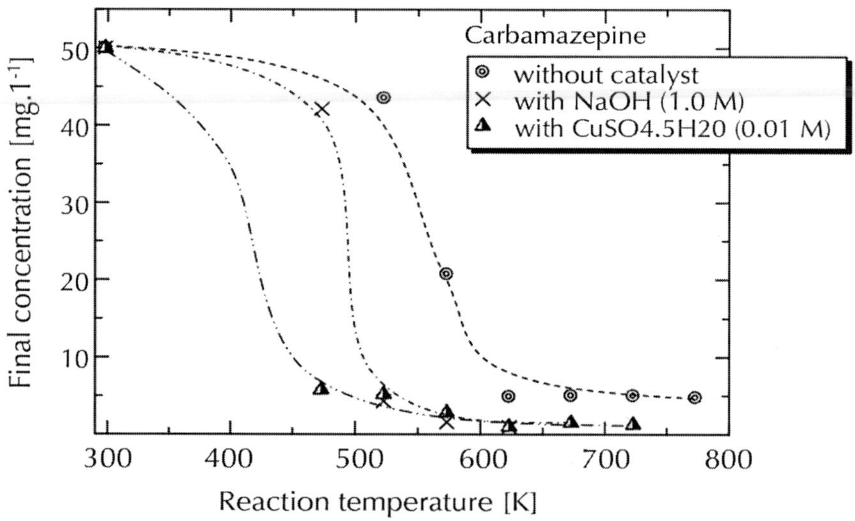

Figure 4: The comparison of final concentration of carbamazepine in different temperatures with catalyst (NaOH and $CuSO_4{\cdot}5H_2O$) and without catalyst.

Catalytic Effect of $CuSO_4$•$5H_2O$ on Metoprolol Destruction

Fig. 5 depicts the effect of temperature on metoprolol destruction efficiency in the presence and absence of $CuSO_4 \cdot 5H_2O$ in a reaction time of 30 min. In the catalyst-free conditions, metoprolol acts as a refractory material under hydrothermal condition (and even at supercritical state of water), while in the presence of $CuSO_4 \cdot 5H_2O$, destruction efficiency reached 100% at 673 K (supercritical state). However, to avoid the limitations of supercritical state and to optimize the energy consumption and destruction efficiency, temperature of 623 K (92% efficiency) could be used.

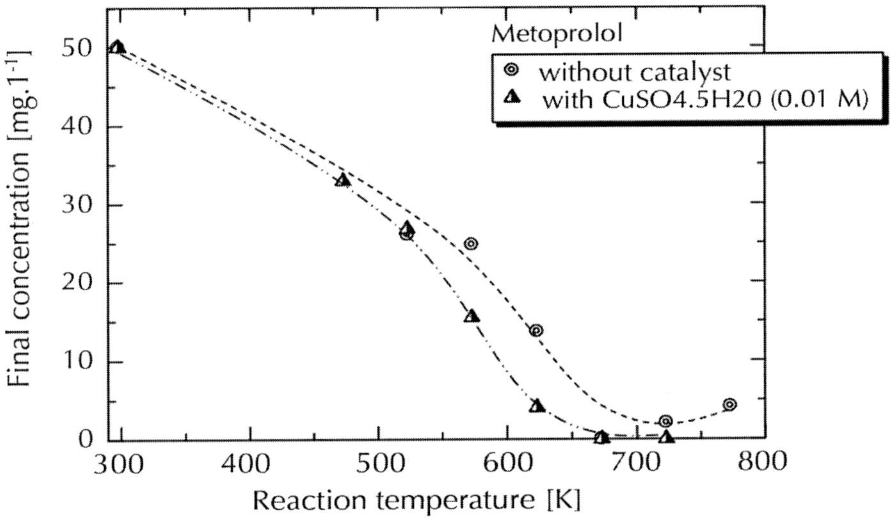

Figure 5: The comparison of final concentration of metoprolol in different temperatures in presence and absence of $CuSO_4 \cdot 5H_2O$.

Catalytic Effect of $CuSO_4$•$5H_2O$ on Sulfamethoxazole Destruction

Fig. 6 presents the final concentration of sulfamethoxazole with and without $CuSO_4 \cdot 5H_2O$ at different reaction temperatures. Results indicate that at low temperatures below 550 K, $CuSO_4 \cdot 5H_2O$ enhanced the destruction efficiency; however, at high temperatures, the presence of catalyst does not have any advantages over catalyst-free conditions. The optimum destruction conditions were in 573 K, for 30 min and without catalyst (99% efficiency), which may be due to the breakdown of sulfur bond on sub-critical water condition.

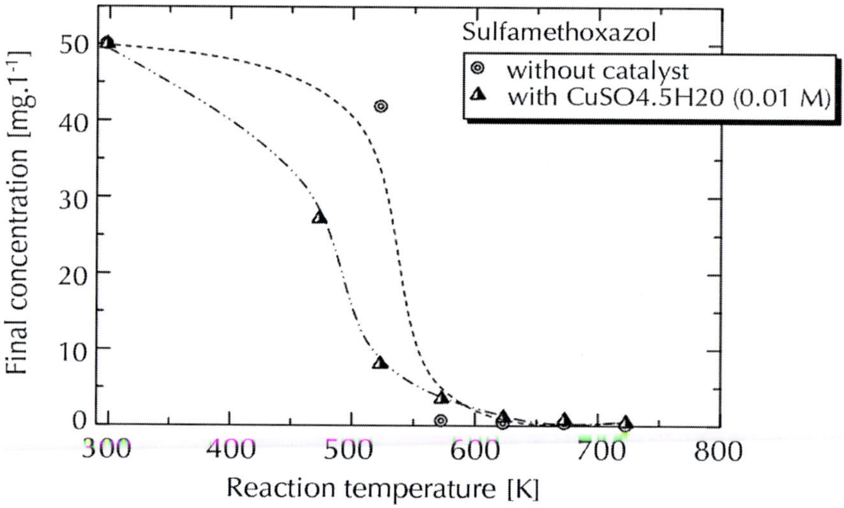

Figure 6: The comparison of final concentration of sulfamethoxazole in different temperatures without catalyst and in the presence of $CuSO_4 \cdot 5H_2O$.

The effect of $CuSO_4 \cdot 5H_2O$ catalyst in 0.01 M on final concentration of the 3 tested pharmaceutical compounds is plotted for comparison in Fig. 7. As shown, carbamazepine was affected more than sulfamethoxazole and metoprolol was affected the least under the optimum (for all compounds) temperature of 623 K (sub-critical water) (with destruction efficiencies of 98%, 98% and 92%, respectively).

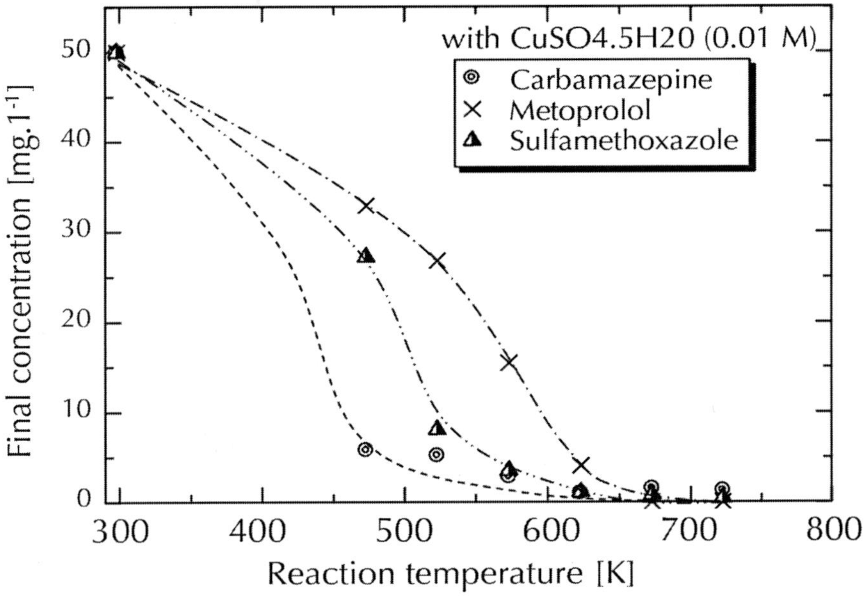

Figure 7: The effects of temperature on final concentration of pharmaceutical compounds at 30 min in the presence of 0.01 M $CuSO4 \cdot 5H2O$ as a catalyst.

Main Reaction Pathways

Proposed reaction pathways for carbamazepine, metoprolol and sulfamethoxazole destruction are shown in Fig. 8, Fig. 9 and Fig. 10, respectively. The degradation occurs in three steps:

Step 1: 200 °C < T < 250 °C: hydrolytic degradation

Step 2: 250 °C < T < 350 °C: water addition to aromatic (and/or) aliphatic bonds

Step 3: T > 350 °C: free radical degradation of saturated aliphatic alcohols [47].

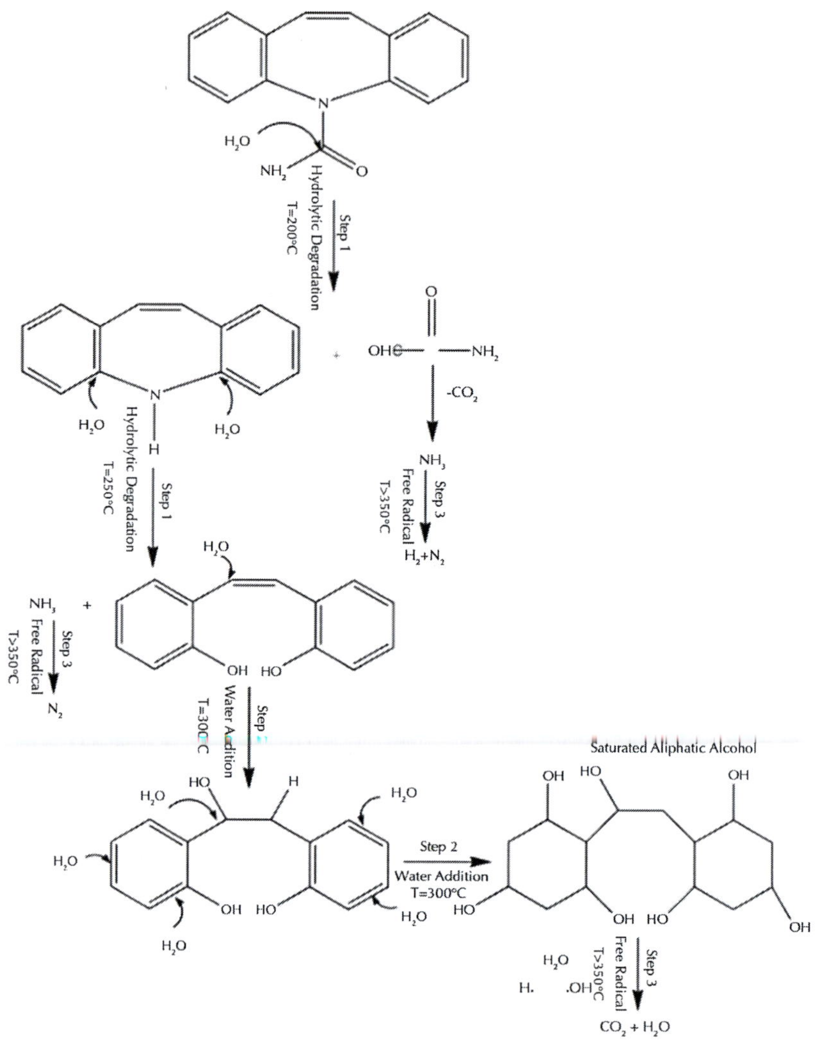

Figure 8: Proposed pathway for carbamazepine degradation in sub-and supercritical water.

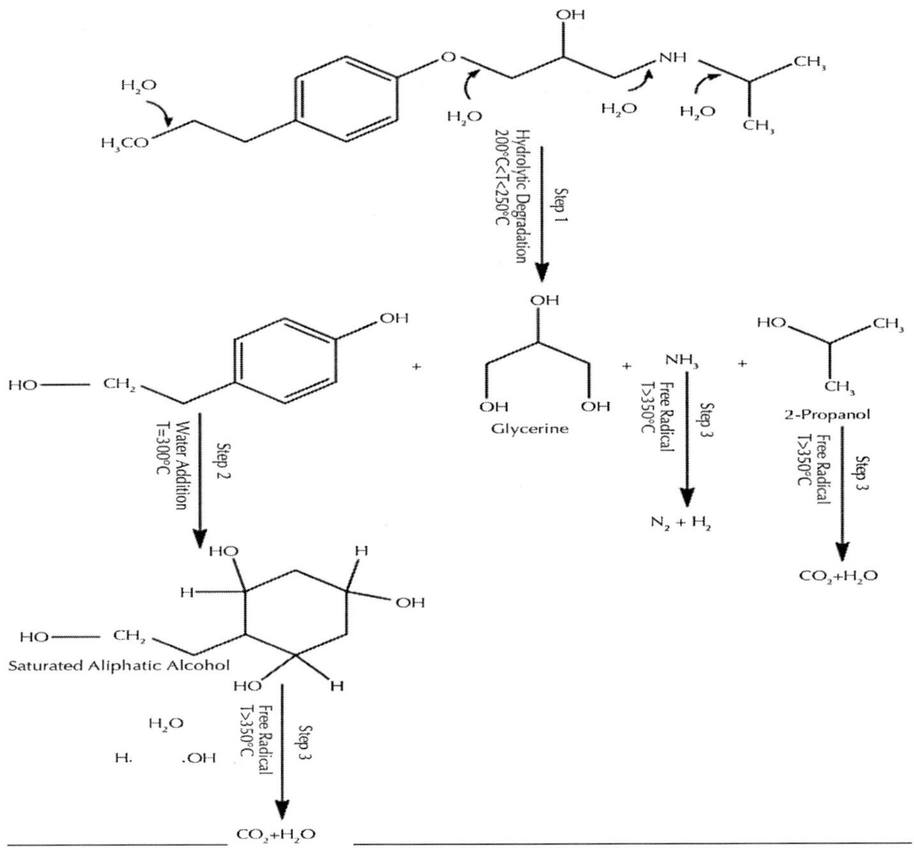

Figure 9: Proposed pathway for metoprolol degradation in sub-and supercritical water.

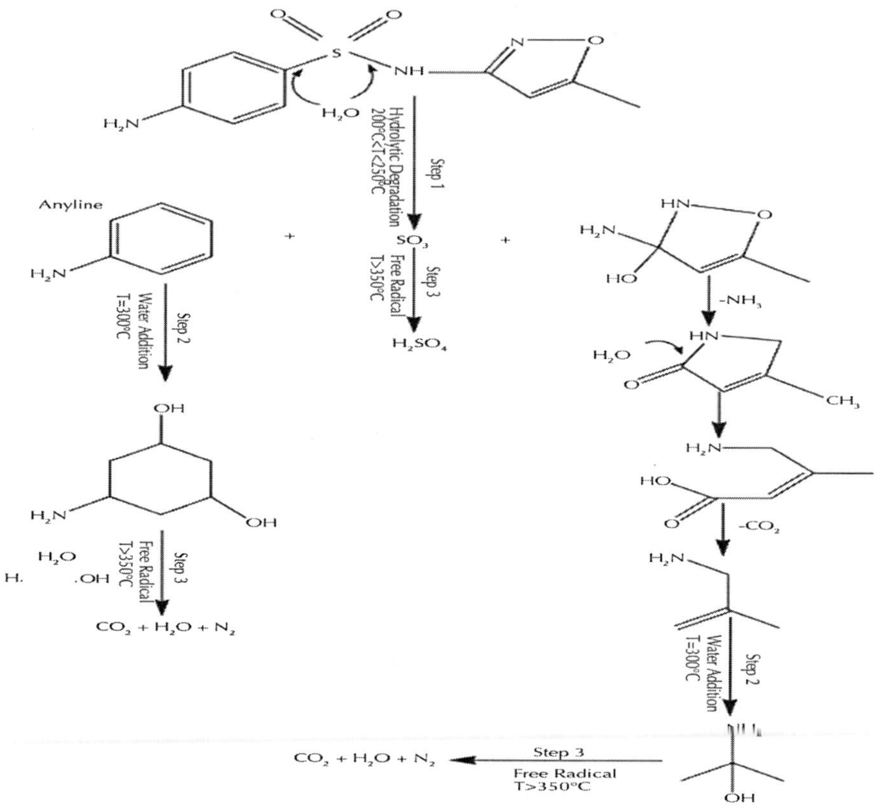

Figure 10: Proposed pathway for sulfamethoxazole degradation in sub- and supercritical water.

In the temperature range of conducted experiments, the reaction products are saturated aliphatic alcohols, CO_2, H_2O and N_2 which do not possess any reported harmful and toxic effects on the environment.

Environmental Benefits of Applying Sub- and Supercritical Water Technology

Most WWTPs are based on activated sludge processes where microorganisms are utilized to convert the pollutants to water and carbon dioxide, or degrade them to acceptable forms. Removal of pollutants from water can also be achieved by stripping into air or by sorption onto sludge with regular discharge [6]. The adverse environmental effects of carbamazepine in low concentrations (1–50 μg/L) are very negotiable; however, ecotoxicological studies have shown reproduction toxicity, decreased enzymatic activity and bioaccumulation in different test organisms. In concentrations above 30.6 mg/L of carbamazepine and 12.6 mg/L of metoprolol, growth retardation and tail deformation of fish embryo was observed [48]. In addition, metoprolol presentation in aquatic environments can accelerate the heart beat rate in Daphnia magna and it has toxic effects on algae and daphina [49] and [50].

Sulfamethoxazole has been found in μg/L concentrations in waste and surface waters. Although usually low levels of antibiotics are seen, their continuous discharge from WWTPs raises concern about their potential adverse ecological effects and their contribution to the development of microbial drug resistance that might impact human health [9].

It can be concluded that even low concentrations of pharmaceutical compounds discharged from WWTPs are important issues to be considered for environmental safety. The comparison between conventional WWTPs and sub- and supercritical water technology removal efficiencies, as presented briefly in Table 1, shows that sub- and supercritical water technology has the highest removal efficiency for carbamazepine, metoprolol and sulfamethoxazole. Additionally, as discussed in Section 3.4, the products of water treatment using this technology are environmental-friendly compounds.

Table 1: Comparison of the removal efficiency of carbamazepine, metoprolol and sulfamethoxazole between conventional wastewater treatment plants and sub- & supercritical water technology

Conventional treatment methods			Sub- and supercritical technology (This research)			
Pharmaceutical compound	**Removal efficiency (%)**	**Ref.**	**T(°C)**	**Time (min)**	**Presence of $CuSO_4{\cdot}5H_2O$**	**Removal efficiency (%)**
Carbamazepine	31	[3] 2011	350	30	No	90.27
	23.1	[2] 2011				
	22	[1] 2011	350	20	No	90.7
	40	[4] 2009	350	30	Yes	98
	10	[6] 2008				
Metoprolol	96	[3] 2011	400	30	No	99.99
	23	[2] 2011				
	32	[1] 2011	400	20	No	96
	66	[4] 2009	400	30	Yes	99.99
Sulfamethoxazole	51.9	[2] 2011	300	30	No	98.84
	44	[7] 2009	300	20	No	96.39
	59	[46]2011				
	86	[51]2011	350	30	Yes	97.81

CONCLUSIONS

In this study, the application of subcritical and supercritical water as an alternative for pharmaceutical wastewater treatment was investigated. Experiments were conducted on three pharmaceutical compounds of carbamazepine, metoprolol and sulfamethoxazole in the temperature range of 473–723 K and reaction time of 5–50 min with and without addition of catalyst (NaOH and $CuSO_4{\cdot}5H_2O$) in the batch reactor. Results show that destruction of selected compounds will increase by increasing temperature and time. Destruction efficiency of 90.7% (623 K) for carbamazepine, 96% (673 K) for metoprolol and 96.39% (573 K) for sulfamethoxazole was gained in catalyst-free conditions. Using $CuSO_4{\cdot}5H_2O$ for carbamazepine is preferred over NaOH, since $CuSO_4·5H_2O$ is capable of increaseing the destruction of carbamazepine at lower temperatures and plus, NaOH causes severe corrosion in the batch reactor. In addition, the presence of $CuSO_4·5H_2O$ enhanced the conversion of carbamazepine and metoprolol; however, it decreased sulfamethoxazole's conversion. Conversions of 98% at 350 °C for 30 min for carbamazepine, 99.99% at 400 °C for 30 min for metoprolol and 97.81% at 350 °C for 30 min for sulfamethoxazole were gained by the addition of $CuSO_4·5H_2O$ as a catalyst. Table 1 compares removal efficiencies gained for these pharmaceutical compounds by conventional treatment methods with sub- and supercritical water technology. The proposed pathway for degradation of above mentioned pharmaceutical compounds illustrates that the products of applying sub- and supercritical water technology are environmentally friendly compounds and do not pose any risks to the environment from an eco-toxicological viewpoint.

In summary, based on the obtained results in this study, subcritical and supercritical water technology can be used as a green alternative in wastewater treatment plants for complete removal of pharmaceutical compounds.

REFERENCES

1. A. Jelic, M. Gros, A. Ginebreda, R. Cespedes-Sanchez, ´ F. Ventura, M. Petrovic, Occurrence, partition and removal of pharmaceuticals in sewage water and sludge during wastewater treatment, Water Research 45 (2011) 1165–1176.
2. S.K.Behera,H.W.Kim,J.-E. Oh,H.-S. Park, Occurrence andremoval of antibiotics, hormones and several other pharmaceuticals in wastewater treatment plants of the largest industrial city of Korea, Science of the Total Environment 409 (2011) 4351–4360.

3. M. Huerta-Fontela, M.T. Galceran, F. Ventura, Occurrence and removal of pharmaceuticals and hormones through drinking water treatment, Water Research 45 (2011) 1432–1442.
4. A.Y.-C. Lin, T.-H. Yu, S.K. Lateef, Removal of pharmaceuticals in secondary wastewater treatment processes in Taiwan, J. Hazardous Materials 167 (2009) 1163–1169.
5. A.Y.-C. Lin, T.-H. Yu, C.-F. Lin, Pharmaceutical contamination in residential, industrial, and agricultural waste streams: Risk to aqueous environments in Taiwan, Chemosphere 74 (2008) 131–141.
6. Y. Zhang, S.-U. Geißen, C. Gal, Carbamazepine and diclofenac: removal in wastewater treatment plants and occurrence in water bodies, Chemosphere 73 (2008) 1151–1161.
7. T. Heberer, Occurrence, fate, and removal of pharmaceutical residues in the aquatic environment: a review of recent research data, Toxicology Letters 131 (2002) 5–17.
8. F. Sacher, F.T. Lange, H.-J. Brauch, I. Blankenhorn, Pharmaceuticals in groundwaters. Analytical methods and results of a monitoring program in Baden-Wu¨ ruttemberg, Germany, J. Chromatography 938 (2001) 199–210.
9. S. Kim, D.S. Aga, Potential ecological and human health impacts of antibiotics and antibiotic-resistant bacteria from wastewater treatment plants, J. Toxicology and Environmental Health, B: Critical Reviews 10 (2007) 559–573.
10. H. Weing rtner, E.U. Franck, Supercritical water as a solvent, Angewandte Chemie International Edition England 44 (2005) 2672–2692.
11. A. Kruse, E. Dinjus, Hot compressed water as reaction medium and reactant: properties and synthesis reactions, J. Supercritical Fluids 39 (2007) 362–380.
12. E.U. Franck, S. Rosenzweig, M. Christoforakos, Calculation of the dielectric constant of water to 1000 ◦C and very high pressures, Berichte der Bunsengesellschaft¨ fur Physikalische Chemie 94 (1990) 199–203.

13. A.G. Carr, R. Mammucari, N.R. Foster, A review of subcritical water as a solvent and its utilisation for the processing of hydrophobic organic compounds, Chemical Engineering J. 172 (2011) 1–17.

14. T. Hirth, R. Schweppe, S. Jahnke, G. Bunte, N. Eisenreich, H. Krause, Degradation processes in sub- and supercritical water, High Pressure Chemical Engineering 12 (1996) 163–168.

15. G. Brunner, Near critical and supercritical water. Part I. Hydrolytic and hydrothermal processes, J. Supercritical Fluids 47 (2009) 373–381.

16. M. Hodes, P.A. Marrone, G.T. Hong, K.A. Smith, J.W. Tester, Salt precipitation and scale control in supercritical water oxidation—Part A: fundamentals and research, J. Supercritical Fluids 29 (2004) 265–288.

17. P. Kritzer, E. Dinjus, An assessment of supercritical water oxidation &LPKT;SCWO&RPKT;: existing problems, possible solutions and newreactor concepts, Chemical Engineering J. 83 (2001) 207–214.

18. P.E. Savage, J. Yu, N. Stylski, E.E. Brock, Kinetics and mechanism of methane oxidation in supercritical water, J. Supercritical Fluids 12 (1998) 141–153.

19. Ph.E. Savage, J. Rovira, N. Stylski, Ch.J. Martino, Oxidation kinetics for methane/methanol mixtures in supercritical water, J. Supercritical Fluids 17 (2000) 155–170.

20. T. Sato, M. Watanabe, R.L. Smith Jr., T. Adschiri, K. Arai, Analysis of the density effect on partial oxidation of methane in supercriticalwater, J. Supercritical Fluids 28 (2004) 69–77.

21. F.Vogel,J.L. DiNaro Blanchard, Ph.A.Marrone, S.F.Rice, P.A.Webley,W.A. Peters, K.A. Smith, J.W. Tester, Review: critical review of kinetic data for the oxidation of methanol in supercritical water, J. Supercritical Fluids 34 (2005) 249–286.

22. G.J. DiLeo, Ph.E. Savage, Catalysis during methanol gasification in supercritical water, J. Supercritical Fluids 39 (2006) 228–232.

23. J. Portella, D. Mateos, F. Mancini, Ch. Marraud, F. Cansell, Hydrothermal oxidation with multi-injection of oxygen: simulation and experimental data, J. Supercritical Fluids 40 (2007) 258–262.
24. E.E. Brock, Y. Oshima, P.E. Savage, J.R. Barker, Kinetics and mechanism of methanol oxidation in supercritical water, J. Physical Chemistry 100 (1996) 15834–15842.
25. G. Anitescu, Z. Zhang, L.L. Tavlarides, A kinetic study of methanol oxidation in supercritical water, Industrial Engineering Chemistry Research 38 (1999) 2231–2237.
26. K. Hirosaka, K. Koido, M. Fukayama, K. Ouryoji, T. Hasegawa, Experimental and numerical study of ethanol oxidation in sub-critical water,J. Supercritical Fluids 44 (2008) 347–355.
27. J. Schanzenbacher, J.D. Taylor, J.W. Tester, Ethanol oxidation and hydrolysis rates in supercritical water, J. Supercritical Fluids 22 (2002) 139–147.
28. U. Armbruster, A. Martin, A. Krepel, Partial oxidation of propane in sub- and supercritical water, J. Supercritical Fluids 21 (2001) 233–243.
29. W.R. Killilea, K.C. Swallow, G.T. Hong, The fate of nitrogen in supercritical-water oxidation, J. Supercritical Fluids 5 (1992) 72–78.
30. Th.D. Thornton, Ph.E. Savage, Phenol oxidation in supercritical water, J. Supercritical Fluids 3 (1990) 240–248.
31. A. Lee, T. Nunoura, Y. Matsumura, K. Yamamoto, Comparison of the effects of theaddition of NaOH on the decomposition of 2-chlorophenol and phenol in supercriticalwater and under supercritical water oxidation conditions, J. Supercritical Fluids 24 (2002) 239–250.
32. I.V. Pérez, S. Rogaka, R. Branion, Supercritical water oxidation of phenol and 2,4-dinitrophenol, J. Supercritical Fluids 30 (2004) 71–87.
33. M. Krajnc, J. Levec, On the kinetics of phenol oxidation in supercritical water, AIChE J. 42 (1996) 1977–1984.

34. P.E. Savage, S. Gopalan, T.I.Mizan,C.J.Martino, E.E.Brock,Reactions at supercritical conditions: applications and fundamentals, AIChE J. 41 (1995) 1723–1778.
35. R. Weber, Sh. Yoshida, K. Miva, PCB destruction in subcritical andsupercritical water evaluation of PCDF formation and initial steps of degradation mechanisms, Environmenral Science Technology 36 (2002) 1839–1844.
36. G. Anitescu, L.L. Tavlarides, Oxidation of biphenyl in supercritical water: reaction kinetics, key pathways, and main products, Industrial Engineering Chemistry Research 44 (2005) 1226–1232.
37. K. Kima, K.S. Kim, S.H. Son, J. Cho, Y.-Ch. Kim, Supercritical water oxidation of transformer oil contaminated withPCBs—a road to commercial plant from bench-scale facility, J. Supercritical Fluids 58 (2011) 121–130.
38. G. Anitescu, V. Munteanu, L.L. Tavlarides, Co-oxidation effects of methanol and benzene on the decompositionof 4-chlorobiphenyl in supercritical water, J. Supercritical Fluids 33 (2005) 139–147.
39. V. Marulanda, G. Bola˜nos, Supercritical water oxidation of a heavily PCBcontaminated mineral transformer oil: laboratory-scale data and economic assessment, J. Supercritical Fluids 54 (2010) 258–265.
40. K.M. Benjamin, Ph.E. Savage, Supercritical water oxidation of methylamine, Industrial Engineering Chemistry Research 44 (2005) 5318–5324.
41. B. Veriansyah, J.-D. Kim, J.-Ch. Lee, Destruction of chemical agent simulants in a supercritical water oxidation bench-scale reactor, J. Hazardous Materials 147 (2007) 8–14.
42. N. Liu, H.-y. Cui, D. Yao, Decomposition and oxidation of sodium 3,5,6- trichloropyridin-2-ol in sub- and supercritical water, Process Safety and Environmental Protection 87 (2009) 387–394.

43. N. Crain, S. Tebbal, L. Li, E.F. Gloyna, Kinetics and reaction pathways of pyridine oxidation in supercritical water, Industrial Engineering Chemistry Research 32 (1993) 2259–2268.
44. B. Veriansyah, T.-J. Park, J.S. Lim, Y.-W. Lee, Supercriticalwater oxidationof wastewater from LCD manufacturing process: kinetic and formationof chromium oxide nanoparticles, J. Supercritical Fluids 34 (2005) 51–61.
45. Z. Fang, S.K. Xu, R.L. Smith Jr., K. Arai, J.A. Kozinski, Destruction ofdecachlorobiphenyl in supercritical water under oxidizing conditions with and without Na2CO3, J. Supercritical Fluids 33 (2005) 247–258.
46. T.-H. Yu, A.Y.-C. Lin, S.C. Panchangam, P.-K.A. Hong, P.-Y. Yang, C.-F. Lin, Biodegradation and bio-sorption of antibiotics and non-steroidalantiinflammatory drugs using immobilized cell process, Chemosphere 84 (2011) 1216–1222.
47. J.L. DiNaro, J.B. Howard, W.H. Green, J.W. Tester, J.W. Bozzelli, Elementary reaction mechanism for benzene oxidation in supercritical water, J. Physical Chemistry A 104 (2000) 10576–10586.
48. E.-J. VandenBrandhof, M. Montforts, Fish embryo toxicity of carbamazepine, diclofenac and metoprolol, Ecotoxicology and Environmental Safety 73 (2010) 1862–1866.
49. M.Cleuvers,Initialriskassessmentforthree&LPAR;beta&RPAR;-blockers found in the aquatic environment, Chemosphere 59 (2005) 199–205.
50. A. Villegas-Navarro, E. Rosas-L, J.L. Reyes, The heart of Daphnia magna: effects of four cardioactive drugs, Comparative Biochemistry and Physiology, C: Toxicology & Pharmacology 136 (2003) 127–134.
51. B. Xu, D. Maob, Y. Luo, L. Xu, Sulfamethoxazole biodegradation and biotransformation in the water–sediment system of a natural river, Bioresource Technology 102 (2011) 7069–7076.

Chapter 6

Development of a Tool, Using CFD, for the Assessment of the Disinfection Process by Ozonation in Industrial Scale Drinking Water Treatment Plants

S. Talvy[a], F. Debaste[a], L. Martinelli[a], E. Chauveheid[b], and B. Haut[a]

[a]Transfers, Interfaces and Processes (TIPs)—Chemical Engineering Unit, Université Libre de Bruxelles, av. F.D. Roosevelt 50, CP 165/67, 1050 Brussels, Belgium

[b]Vivaqua, rue aux Laines 70, 1000 Brussels, Belgium

ABSTRACT

Foreseen standards regarding microorganism content for drinking water require assessment of the capability of existing plants to reach the upcoming requirements. This paper presents the development of a tool to assess this capability in a commonly encountered key step of water disinfection: ozonation. In this paper, this tool is applied to the test case of an ozonation channel of the Belgian drinking water producer Vivaqua. This tool is based on a mathematical model of the momentum and mass transport phenomena in an ozonation channel. The gas–liquid flow is coupled to ozone mass transfer and kinetics describing the ozone and microorganisms concentrations decay. The degradation of Bacillus subtilis spores, as a representative of resistant microorganisms, is implemented in the model. The model takes explicitly into account the bubble size variation and its impact on mass transfer. Bubbles sizes and kinetics parameters are estimated based on dedicated experiments. The model is partially validated by comparing simulations results, obtained using computational fluid dynamics, to experimental residence time distributions, residual ozone concentration and Bacillus subtilis spores degradation efficiency measurements obtained on the studied ozonation channel. It is shown that, at the industrial scale, bubble diameter variation has a significant impact on ozone concentration in the liquid at the reactor exit. Using the tool, it is also shown that, the ozonation channel of Vivaqua can be used to achieve degradation of resistant microorganisms but only with its maximal flow rate and concentration of ozone injection. Moreover, at low operating temperature, some microorganisms that present latency towards reaction with dissolved ozone might hardly be destroyed.

INTRODUCTION

Nowadays, according to the European Union Council Directive 98/83/EC (1998), drinking water must be devoid of Escherichia coli and Enteroccocus at the production plant outlet. However, in the coming years stricter European standards are foreseen. These new standards will include the inactivation of more resistant microorganisms like Cryptosporidium and the ability of the process to neutralise peak pollution events. Therefore, there is a need to assess the efficiency of ozonation channels in case of harsher functioning conditions as high concentration of microorganisms or low temperatures.

Various phenomena, such as liquid and gas flow, gas–liquid mass transfer and microorganism degradation have to be taken into account to obtain an efficient treatment installation. The key parameters of all those phenomena (flow pattern, mass transfer coefficient, degradation kinetics ...) are interconnected and should be studied following an integrated approach.

A mathematical model of transport phenomena, implemented in Computational Fluid Dynamics (CFD), is a convenient tool for flow and mass transport analysis that can integrate explicitly simplified models of other phenomena such as the disinfection. Former works have shown that the use of CFD is efficient for the characterisation and the optimisation of unit operations in industrial water treatment plants, as ozonation channel or aeration tank (Do-Quang et al., 1999, Cockx et al., 1999, Fayolle et al., 2007 and Wols et al., 2010). Some studies (Cockx et al., 1999, Zhang, 2006, Audenaert et al., 2010 and Wols et al., 2010) addressed specifically the characterisation of disinfection in ozonation process at industrial-scale either with CFD or global modelling. Other authors focused, at pilot-scale, on modelling ozone gas–liquid transfer (Gong et al., 2007) or disinfection (van der Helm et al., 2008 and Bartrand et al., 2009) for instance.

The ozone exposure (CT) is the integral with respect to time of the ozone concentration seen by microorganisms, during their contact with ozone in an industrial plant (Cockx et al., 1999). Cockx et al. (1999) highlighted increases of the CT in an industrial ozonation channel by changing ozone injection position and adding more baffles in the channel. Zhang (2006) realised comparisons between different models for gas phase transport and included disinfection and bromate formation mechanisms in a full-scale water treatment plant. The author also aims at improving the CT prediction. Kim (2007) also accounts for disinfection and bromate formation, thanks to Ozone Contactor Model software to provide recommendations for design improvement. Audenaert et al. (2010) developed a global model for scenario analysis regarding disinfection and bromate formation. Wols et al. (2010) developed different calculation methods so as to predict the disinfection in ozone systems also in a view to calculate the CT.

This prospective research is in the continuity of the previously quoted articles. The objective of this work is to develop a tool that can be used to assess the disinfection by ozonation in drinking water treatment plant, in view of the optimisation of the process and the anticipation of the new European standards. This tool can be used in risk analysis studies by predicting residual ozone concentration and microorganisms degradation.

This tool is based on a mathematical model, fed by laboratory experiments and implemented in CFD. The model describes the ozonation channel in terms of gas–liquid flow, ozone transfer, transport, reaction and disinfection.

The tool is applied to the test case of an ozonation channel of the Belgian drinking water producer Vivaqua. Nowadays, this existing installation is used all year round for the successful elimination of Escherichia coliand Enteroccocus. From Vivaqua measurements (Chauveheid, 2004), this plant is known to work without generating problematic concentrations of bromates considering the European Union Council Directive 98/83/EC (1998), and to contain a negligible amount of compounds leading to a Fast Ozone Demand (FOD). The measurements from Vivaqua confirmed that differential values of UV254 nm between the inlet and outlet of the ozonation channel are in average equal to 0.6 m^{-1}. This confirms that the FOD must be lower than 0.6 m^{-1}, thus considered negligible.

However, nothing guaranties that the elimination of more resistant microorganisms is efficient enough in regard of the upcoming legislation. In this paper, the developed tool is used to assess the elimination, in the Vivaqua ozonation channel, of Bacillus subtilis spores, considered as a representative for resistant microorganisms such as Giarda or Cryptosporidium, both more difficult to measure. The model parameters values are identified from dedicated experiments (decomposition of ozone and disinfection kinetics, bubbles diameter) on water coming from the Vivaqua plant or from literature (mass transfer coefficient, ozone solubility ratio in water, ozone diffusivity coefficient in water).

The originality of this work lies in the implementation, in the model, of the bubble diameter variation with respect to the hydrostatic pressure variation and an adapted disinfection law accounting for Bacillus subtilisspores resistance to ozonation, taken into account in the model through the introduction of a latency time in the disinfection kinetics.

This paper focuses on presenting the developed tool and examples of its use regarding the Vivaqua ozonation channel. The materials and methods section begins with the description of the Vivaqua plant, in which the considered ozonation channel is integrated. Then, the tool construction is described, beginning by the presentation of the model equations. The experiments used to identify the model parameters values follow. Experimental Residence Time Distribution (RTD) measurements, used to partially validate the model are then presented. The description of the tool construction is finished by presenting the numerical CFD methodology used. In the results and discussion part, the partial validation of the model using RTD and residual ozone and inactivation efficiency measurements on the Vivaqua ozonation channel is first presented. Then, in order to illustrate the use of the tool, different parameters values variations are tested in order to simulate different practical scenarios. Conclusions concerning both the tested scenarios and further developments of the tool end this paper.

MATERIALS AND METHODS

Tailfer Drinking Water Plant

A large part of the drinking water distributed in the region of Brussels capital, Belgium is produced in the Tailfer plant which is managed by the water treatment company Vivaqua. This plant has a production capacity of 240,000 m^3 day^{-1}.

The water treatment plant is presented in Fig. 1. Raw water is pumped in the Meuse River (Belgium). After a succession of pre-treatments and operations of physical separation, the water is finally disinfected by ozonation. After the disinfection step, a post-filtration, chlorination and a pH correction steps are operated. The drinking water is then distributed to Brussels. An ozonation module is divided into two identical ozonation channels. A 3D representation of an ozonation channel is shown in Fig. 2. The compartment including 18 spargers is called chamber 1, the one next to its right is called chamber 2. Ozone enriched air is produced in situ by corona discharge using a Ozonia ozone generator (Ondeo Degremont, S.A., France).

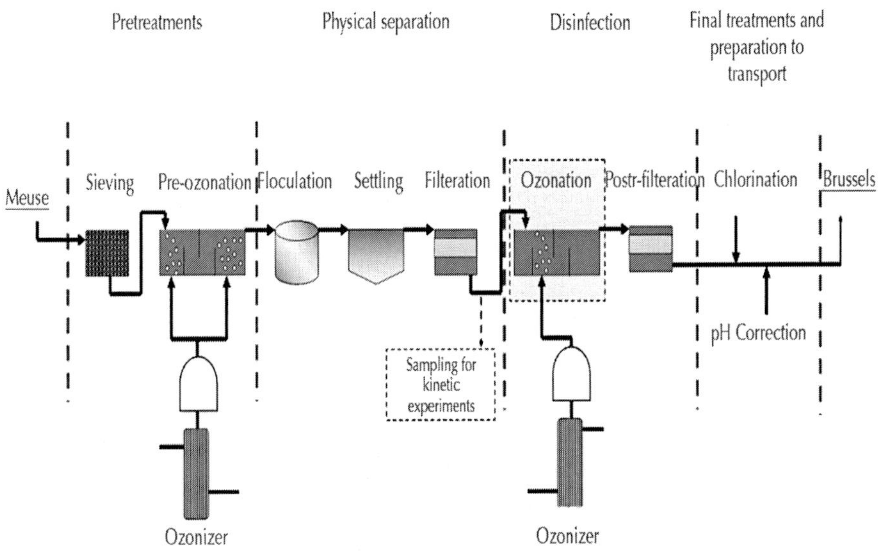

Figure 1: Vivaqua drinking water production plant representation.

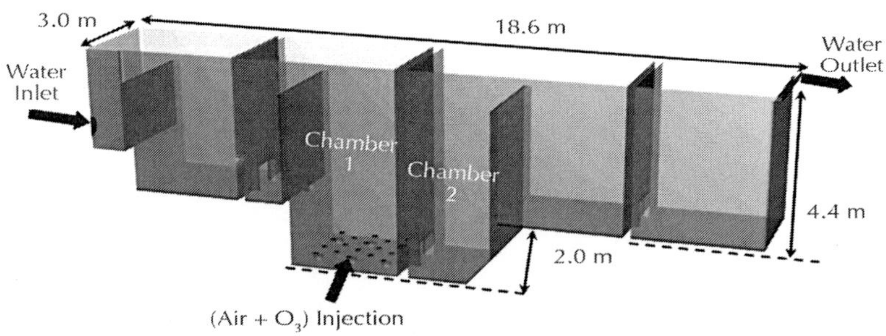

Figure 2: The 3D presentation of an ozonation channel of the Tailfer drinking water plant.

Considering an ozonation module, the applied water flow rate ranges within 1000 m^3 and 2900 m^3 h^{-1}. Inlet ozone concentration in the gas phase is normally set to 20 g Nm^{-3}. Good practice implies that the dissolved ozone concentration has to be larger than a threshold of 0.4 mg l^{-1} at the outlet of the module. When needed, the inlet ozone concentration in the gas can be raised up to 40 g

Nm^{-3} in order to remain above this threshold. If this operation does not satisfy the threshold, the gas flow rate is switched from 48 to 72 $Nm^3\ h^{-1}$, keeping the concentration at 40 g Nm^{-3}.

Model Equations

In this section, the model equations regarding the gas-liquid flow and the various reaction rates involving dissolved ozone are presented.

Two-phase Flow Model

A classical eulerian isothermal two-fluid model is developed, similar to the one used in Talvy et al. (2007a) and Cockx et al. (1999) for example. The drag coefficient is also evaluated as in Talvy et al. (2007a).

The bubble radius r_b (m) is evaluated locally using the following equation, based on the ideal gas law:

$$r_b = r_{b0}\left(\frac{P_{atm}+\rho^L gH}{P_{atm}+\rho^L g(H-z_b)}\right)^{1/3} \tag{1}$$

where r_{b0} (m) is the bubble radius at the injection, P_{atm} (kg m^{-1} s^{-2}) is the atmospheric pressure, ρ^L(kg m^{-3}) is the liquid volumetric mass, g (m s^{-2}) is the gravity acceleration, H (m) is the depth of the injector and z_b (m) is the vertical upward position of the considered bubble with z_b=0 at the injector. The variation of the bubble diameter affects its volume and thus the ozone concentration in the gas.

The bubble volumetric mass ρ^G (kg m^{-3}) is also calculated according to the local hydrostatic pressure:

$$\rho^G = \frac{M\left[P_{atm} + \rho^L g(H - z_b)\right]}{R\theta} \tag{2}$$

where M is the air molar mass (kg mol^{-1}), R (8.314 J mol^{-1} K^{-1}) is the ideal gas constant, θ(K) is the temperature. Ozone molar fraction in Eq. (2) is neglected since it is smaller than 2% in the gas phase in the channel.

Ozone Transfer-reaction and Disinfection Model Equations

The transport equation for the ozone in the gas phase is written as

$$\partial_t\left(\alpha^G \rho^G C^G_{O_3}\right) + \partial_i\left(\alpha^G \rho^G C^G_{O_3} u^G_i\right) = -k^L a\left(m\, \rho^G C^G_{O_3} - \rho^L C^L_{O_3}\right) \tag{3}$$

where $C^G_{O_3}$ (mg kg^{-1} of gas phase) is the local statistical average of the ozone concentration in the gas phase, α^G (m^3 of gas m^3 of device) is the local statistical average of the gas volume fraction, u^G_i (m s^{-1}) is the local statistical average of the ith component of the gas velocity, k^L (m^3 of liquid m^{-2} of interface s^{-1}) is the mass transfer coefficient, a (m^2 of interface m^{-3} of device) is the interfacial area, m (mg m^{-3}water/mg m^{-3} gas) is the ozone solubility ratio in water defined by Morris (1988), $C^L_{O_3}$ (mg kg^{-1} of liquid phase) is the local statistical average of the ozone concentration in the liquid phase.

The bubble interfacial area a (m^2 of interface m^{-3} of device) is expressed as

$$a = 3\frac{\alpha^G}{r_b} \tag{4}$$

The mass transfer coefficient k^L can be estimated by the classical correlation from Calderbank and Moo-Young (1961) (Eq. (5)) for freely rising bubbles whose diameter is higher than 2.5 mm or by the Higbie (1935) model (Eq. (6))

$$k^L = 0.42\left(\frac{(\rho^L-\rho^G)g\mu^L}{\rho^{L2}}\right)^{0.33}\left(\frac{\mu^L}{\rho^L D^L_{O_3}}\right)^{-0.5} \tag{5}$$

$$k^L = \frac{2}{\sqrt{\pi}}\sqrt{\frac{D^L_{O_3}\,G}{2r_b}} \tag{6}$$

where μ^L (kg m^{-1} s^{-1}) is dynamic viscosity of the liquid, $D^L_{O_3}$ (m^2 s^{-1}) is the molecular diffusion coefficient of ozone in the liquid phase (Table 1) defined by Johnson and Davis (1996) and G (m s^{-1}) is the bubble slip velocity. These correlations can be used since, in the Vivaqua ozonation channel, gas volume fraction is very low (around 0.2%— Table 3) and thus compatible with the freely rising bubble assumption. Higbie (1935) correlation was used by Cockx et al. (1999) with bubble of diameters 3 mm, which is similar to what is observed in Vivaqua ozonation channel (see Section 2.3.1). Both correlations consider that bubbles have a mobile interface easily renewable. As the values of kL computed from Eqs.(5) and (6) differ from less than 2% in all the conditions considered in this work, any of those equations can be used. Eq. (5) is used in the rest of this work.

Table 1: Parameters values at 5, 8 and 20 °C

T(°C)	5	8	20
m (–)	0.48	0.44	0.31
$_{DO3}$(m² s⁻¹)	1.62×10^{-9}	1.66×10^{-9}	1.74×10^{-9}
k^L(m s⁻¹)	3.39×10^{-4}	3.46×10^{-4}	3.72×10^{-4}
k_d (s⁻¹)	3.13×10^{-4}	3.59×10^{-4}	6.03×10^{-4}
k_μ (l mg⁻¹ s⁻¹)	0.86×10^{-2}	1.19×10^{-2}	2.90×10^{-2}
k_l (s⁻¹)	0.35×10^{-2}	0.46×10^{-2}	1.30×10^{-2}

The transport and reaction equation of the ozone in the liquid phase is written as

$$\partial_t(\alpha^L\rho^L C^L_{O_3})+\partial_i(\alpha^L\rho^L C^L_{O_3}u^L_i)$$
$$=\partial_i(\alpha^L\rho^L D^{tL}\partial_i C^L_{O_3 3})+k^L a(m\,\rho^G C^G_{O_3}-\rho^L C^L_{O_3})-\alpha^L k_d\rho^L C^L_{O_3} \quad (7)$$

where u^L_i (m s⁻¹) is the local statistical average of the ith component of the liquid velocity, α^L (m³ of liquid m³ of device) is the local statistical average of the liquid volume fraction, D^{tL} (m² s⁻¹) is the turbulent diffusion coefficient in the liquid phase, $_{kd}$ (s⁻¹) is the first order kinetic constant of the ozone decomposition in water. The turbulent diffusion coefficient is assumed to be proportional to the turbulent dynamic viscosity μ^{tL} (kg m⁻¹ s⁻¹)

$$D^{tL}=\frac{\mu^{tL}}{0.7\rho^L} \quad (8)$$

The transport and degradation equation of Bacillus subtilis spores in the liquid phase written as

$$\partial_t(\alpha^L\rho^L N)+\partial_i(\alpha^L\rho^L N u_i^L)=\partial_i(\alpha^L\rho^L D^{tL}\partial_i N)-\alpha^L 10^{-3}\rho^{L2}k_\mu C_{O_3}^L N(1-e^{-k_l t_c}) \quad (9)$$

where N (CFU kg^{-1} of liquid phase) is the number of Bacillus subtilis spores colony-forming unit per kg of liquid phase, k_μ (l mg^{-1} s^{-1}) is the Bacillus subtilis spores degradation kinetic constant, k_l (s^{-1}) is a latency constant related to a resistance time of the spores towards the ozone and t_c (s) is the local contact time of the Bacillus subtilis spores with ozone. The calculation of this contact time is presented in the Section 2.2.3.

Compared to the classical delayed Chick–Watson equation (Renneker et al., 1999), the influence of the local ozone concentration and the latency time on the inactivation kinetics is directly taken into account.

Hydraulic Retention Time, Contact Time and Residence Time Distribution Calculation

The Hydraulic Retention Time (HRT) t_l (s) is evaluated, thanks to the transport equation of a scalar, representing the age of the liquid elements in the ozonation channel. This equation is written as

$$\partial_t\alpha_L t_l+\partial_i(\alpha_L u_i^L t_l)=\partial_i(\alpha^L D^{tL}\partial_i t_l)+\alpha^L\cdot 1 \quad 10)$$

The $+\alpha^L\cdot 1$ term in Eq. (10) is a source term implying that the liquid elements will age a time unit per a unit of time during the iterations. t_l is set to zero at the entrance of the ozonation channel. This approach was previously implemented successfully to determine bubble age by Talvy et al. (2007b) in an airlift loop reactor.

The contact time t_c (s) of Bacillus subtilis spores with ozone is defined as 0 upstream of the entrance of chamber 1 (Fig. 2). Downstream of this entrance, it is defined as

$$t_c = t_l - t_{l1} \tag{11}$$

where t_{l1} (s) is the value of tl at the entrance of chamber 1.

A residence time distribution in the channel is computed by solving a transport equation for a tracer, an inert specie of concentration C^L_{tr} (mg kg^{-1} of liquid phase)

$$\partial_t(\alpha^L \rho^L C^L_{tr}) + \partial_i(\alpha^L \rho^L C^L_{tr} u^L_i) = \partial_i(\alpha^L \rho^L D^{tL} \partial_i C^L_{tr}) \tag{12}$$

The initial concentration of the tracer is equal to zero in the whole domain except in a small cubic volume of 8×10^{-3} m^3 at the top of the chamber 1 (Fig. 2) where the concentration is fixed at a uniform arbitrary value, in order to simulate the tracer experiments (see Section 2.3.4). The residence time distribution is the evolution of the outlet tracer concentration as a function of time.

Bubble Size, Residence Time Distribution and Kinetic Parameters Determination

This section presents the experimental methodologies followed to estimate the value of the parameters required for the model and to obtain experimental RTDs to validate the model. The parameters values to obtain are the bubble diameter at the bottom of the channel and the kinetic constants of the ozone consumption kinetic and of the Bacillus subtilis spores degradation.

Bubble Size Determination

The size distribution of the bubbles produced by a porous sparger having the same characteristics as those present in the Vivaqua ozonation channel has been experimentally evaluated. The sparger is placed at the bottom of a tank filled with water. Bubbles are then produced using an air flow rate equal to the one used, per sparger, in the Vivaqua ozonation channel. Shadows of the bubbles at camera height are obtained by the use of an imaging method using parallel beam backlighting. The experimental device is described in Larcy (2006). The images obtained have been processed by the software described in Dehaeck (2007). This software fits the shadows of the bubbles with ellipses. Assuming that the shape of the bubble presents an axial symmetry around the minor axis of the fitted ellipse, a volume equivalent diameter can be obtained for each bubble. The bubble diameter used in this work is the most probable diameter of a distribution composed of approximately 50 bubbles. The value obtained is 3.3 mm.

Ozone Decomposition Kinetic Parameters

The value of the first order decay kinetic constant k_d (Eq. (7)) has been obtained by dedicated experiments in a thermostatic batch reactor with a volume of 12 l of filtrated water from the Tailfer plant (Fig. 1) (pH=7.3) at three temperatures (5, 15 and 25 °C).

The time evolution of the ozone concentration in the liquid is followed by the Indigo method (Bader and Hoigné, 1981). For a batch reactor, this time evolution can be modelled by the following:

$$C_{O_3}^{L}(t) = C_{O_3,0}^{L} e^{-k_d t} \tag{13}$$

where $C^L_{O_3}(t)$ is the ozone concentration in the reactor at time t and $C^L_{O_3,0}$ is the initial (t=0) ozone concentration in the reactor. For each experimental condition, the values of k_d is estimated by a least-square optimisation of the fitting of $C^L_{O_3}(t)$ from Eq. (13) on the corresponding experimental data.

The values of k_d at different temperatures are obtained thanks to an Arrhenius law

$$k_d(\theta) = k_{d,0} e^{-E_{A,k_d}/R\theta} \tag{14}$$

where $k_{d,0}(s^{-1})$, the pre-exponential constant, and $E_{A,kd}$ (J mol^{-1}), the activation energy, are calculated by a least square-fitting of Eq.(14) on the experimental values of kd as a function of temperature. k_{d0} and are, respectively, equal to 113.6 s^{-1} and 2.9×10^4 J mol^{-1}.

The values of kd used for CFD simulations, at 5, 8 and 20 °C, calculated, thanks to Eq. (14), that are summarised in Table 1.

Based on these values, the Hatta number Ha can be calculated to check whether the reaction of ozone decomposition has an effect on the ozone transfer rate. This number can be written as

$$\mathrm{Ha} = \frac{1}{k^L}\sqrt{D^L_{O_3}\, k_d} \tag{15}$$

Result shows that an acceleration of the gas–liquid transfer by reaction does not need to be accounted for as Ha=0.003 which is lower than the classical threshold value of 0.3 (Beltran, 2004).

Disinfection Kinetics Parameter

The values of k_μ and kl (Eq. (9)) used in this work have been obtained from dedicated batch experiments.

A Bacillus subtilis colony from the river Meuse is isolated and cultivated on a tryptic soy agar medium. This colony is incubated at 30 °C+/−2 °C during 7 days. The colonies are re-suspended in sterilised distilled water. The suspension is placed in an 80 °C water bath for 15 min in order to have sporulating of the bacilli.

For each Bacillus subtilis spore degradation experiment, from 5×10^5 to 10^6Bacillus subtilis spores are injected in the batch reactor described in Section 2.3.2. Different experiments are realised with three ozone concentrations (0.5, 1 and 1.5 mg l^{-1}) and at three temperatures (5, 15 and 25 °C). Microbiological samples are taken in sterile bottles containing thiosulfate in order to quench the ozone residual. The thiosulfate concentration is 5 mg l^{-1} in each sample. In order to quantify the remaining Bacillus subtilis spores, the samples are filtrated through a cellulosic membrane (pore diameter=0.45 µm). After filtration, the membrane is placed on Mannitol Egg-Yolk Polymixin B agar, incubated at 30 °C and the number of Bacillus subtilis colonies is enumerated according to ISO 7932 standard (1994).

For the considered batch reactor, the time evolution of the number of colony-forming unit per kg of liquid phase N (t) can be modelled by the following:

$$\frac{dN(t)}{dt} = -10^{-3}(\rho^L)^2 k_\mu C^L_{O_3,0} \exp(-k_d t) N(t)(1-\exp(k_l t)) \quad (16)$$

For each set of experimental conditions, the values of k_μ and kl are estimated by a least-square optimisation of the fitting of N as a function of t obtained by the integration of Eq. (16) on the corresponding experimental data. Values of k_μ and kl at different temperatures are presented in Table 1. The values are obtained following the same methodology for the evaluation of the values of k_d. For k_μ and kl, the Arrhenius laws are written as

$$k_\mu(\theta) = k_{\mu,0} e^{-E_{A,k_\mu}/R\theta} \quad (17)$$

$$k_l(\theta) = k_{l,0} e^{-E_{A,k_l}/R\theta} \tag{18}$$

where $k_{l'0}$ equals to 5.18×10^8 s^{-1}, $E_{A,kl}$ equals to 5.9×10^4 J mol^{-1}, $k_{\mu'0}$ equals to 3.37×10^7 mg l^{-1} s^{-1}and $E_{A,k\mu}$ equals to 5.1×10^4 J mol^{-1}. The constants k_μ and k_l are adjusted to the experimental values of bacilli concentration. The correlation coefficient R^2 is equal to 0.973.

Residence Time Distribution Measurement and Comparison to Simulations

In prior experiments realised on the Vivaqua ozonation module considered in the paper, radioactive tritium particles were injected at the top of chamber 1 in the ozonation channel and were collected at the end of the ozonation module. The resulting experimental outlet concentration as a function of time is proportional to the RTD in the part of the module downstream to the top of chamber 1.

However, the measurement probe was not located directly at the outlet of the ozonation module but rather after a long sampling tube whose detailed characteristics, regarding dimension and liquid flow rate, are no longer available.

This tube influenced the signal which obtained experimentally preventing direct comparison between simulation and experimental results. Therefore, to take the tube into account, it is proposed to modify the simulated RTD considering that the ozonation channel is followed by a tube where a plug-dispersion flow takes place. The simulated concentration e $C^L_{tr,tube}(t)$ at the exit of this series of two reactors is written as (Levenspiel, 1999)

$$C^L_{tr,tube}(t) = \frac{1}{2}\int_0^t C^L_{tr,channel}(t-t')\sqrt{\frac{Pe}{\pi\tau_{tube}t}}\exp\left[-\frac{Pe(\tau_{tube}-t')^2}{4\tau_{tube}t'}\right]dt' \quad (19)$$

where $C^L_{tr,channel}(t)$ is the simulated tracer concentration at the exit of the ozonation channel, τ_{tube} (s) is the average residence time of the liquid in the sampling tube and Pe (–) is the Peclet number of the tube. τ_{tube} can be estimated based on known experimental values. Indeed, the observed average total residence time τ_{tot} (s), computed based on the experimental RTD, can be expressed as

$$\tau_{tot} = \tau_{channel} + \tau_{tube} \quad (20)$$

where $\tau_{channel}$ (s) is the average residence time in the ozonation channel, directly given by the ratio between the channel volume downstream of the top of chamber 1 and the liquid flow-rate going through it.

For a first liquid flow rate, the value of the Peclet number is estimated by a least-square optimisation of the fitting of $C^L_{tr,tube}(t)$ given by Eq. (19) on the experimental data, using the calculated value of τ_{tube}. Then, the identified value of Pe and the calculated value of τ_{tube} are used to compare simulations and experimental results at a different liquid flow rate in the ozonation channel. Good fitting would mean that the flow modelling is partially cross-validated.

Simulation Methodology and Boundary Conditions

The model equations are solved numerically using CFD. The geometry is built and meshed with Gambit®2.4. The geometry is composed of 644,000 cells. Around the air diffusers the mesh is composed of tetrahedral cells. In the remaining part of the domain, the cells are hexahedral. The CFD simulations are realised with Ansys Fluent® 6.3. The convection terms in the equations are discretised with the QUICK scheme for the momentum and the volume fraction transport equations. The second order UPWIND scheme is used to discretise the transport equations of the turbulence kinetic energy and its dissipation rate. The first order UPWIND scheme is used for the scalar transport equations.

At the gas and liquid inlets, the velocities and concentrations of the different species (ozone, microorganisms) are fixed according to operating conditions. At the liquid inlet, tl is set at zero. At the gas inlet, ozone concentration and bubble radius are fixed. Top boundary condition is a degassing condition (implemented through a user defined function in fluent code, Fluent Inc. (2006)) for the gas and a zero shear stress condition for the liquid. All the other boundaries are treated using standard wall functions. The boundary conditions for the tracer transport equation are zero normal flux conditions for all the boundaries except at the exit where a zero diffusive flux is imposed.

The simulations are run following different successive steps. The next step is considered only after convergence of the current one. The steps are:

- simulation of the gas-liquid flow;
- simulation of the transport of ozone from the gas phase to the liquid phase, transfer and reaction of ozone in the liquid phase;
- Simulation of the disinfection kinetic.

Steps 2 and 3 can be performed successively since the disinfection kinetic does not affect the ozone residual concentration.

RESULTS AND DISCUSSION

In this section, the developed tool is used to compare its prediction with experimental data and to highlight some of the tool potential uses.

Firstly, the simulated RTDs are compared to the experimental ones. Secondly, a case study (S1) for which an experimental value of residual ozone concentration is known is presented and discussed regarding ozone concentration at the exit of the channel and Bacillus subtilis spores degradation. Then a reference case (S2) corresponding to industrial minimal ozone quantity injection is presented and discussed. Then, the impact of the ozone degradation kinetic and of variation of the bubble size with static pressure is discussed by comparing a reference case (S2) to two cases (S3 and S4) simulating the same conditions but with no ozone degradation kinetic and constant bubble size, respectively. Finally, a set of cases of industrial interest (S5–S8) for different operating conditions is presented and discussed.

Residence Time Distribution

Fig. 3 presents the comparison between experimental and simulated RTDs. Fig. 3a shows the comparison in the conditions used for the model fitting, at a liquid flow rate of 2500 $m^3\ h^{-1}$ while Fig. 3b shows the comparison in the conditions used for the cross-validation at a liquid flow rate of 2900 $m^3\ h^{-1}$. On both figures, the simulated tracer concentrations at the exit of the channel $C^L_{tr,channel}(t)$ and after the sampling tube $C^L_{tr,tube}(t)$ are presented. The obtained values of τ_{tube} and Peclet number are 2 min and 6, respectively. The good fitting of $C^L_{tr,tube}(t)$ with the experiments in the two situations with a unique value of Pe tends to show that the liquid flow is correctly predicted. However, the fitting of this Peclet number might hide some errors, present in all the simulations, due to the model assumptions or to the numerical resolution. Therefore, a more detailed validation might be of interest.

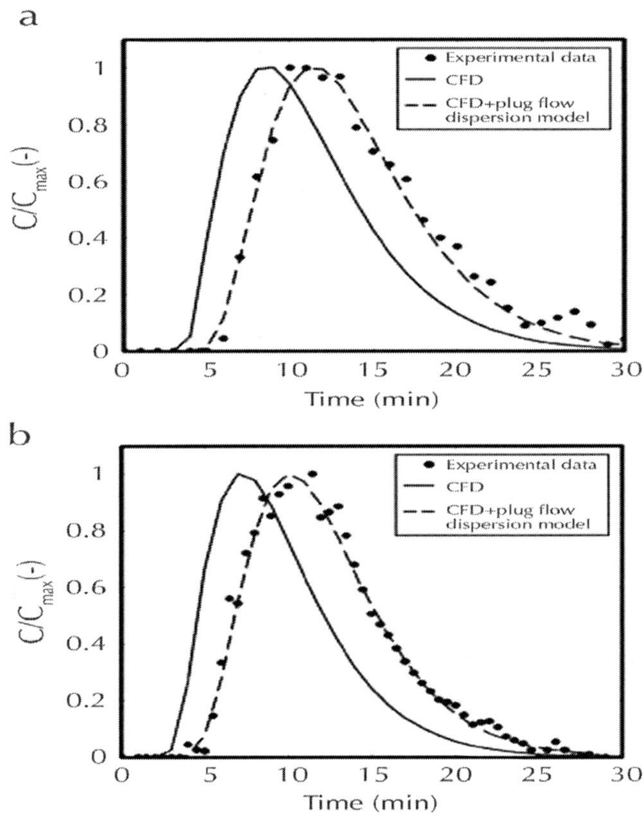

Figure 3: Comparison between modelled and experimental concentration of the tracer as a function of time. $_{Cmax}$ is the highest concentration achieved by a curve. —— CFD RTD signal out of ozonation channel ($C^{L}_{tr,channel}/C_{max}$) — — CFD+plug flow dispersion model ($C^{L}_{tr,tube}/C_{max}$).(a) Liquid flow rate equal to 2500 $m^3\ h^{-1}$ and (b) liquid flow rate equal to 2900 $m^3\ h^{-1}$.

Application of the Tool to an Industrial Case and to a Reference Case

Table 2 summarises, for all the considered cases, the operating conditions, maximum theoretical dissolved ozone concentration, average ozone concentration at chamber 1 outlet, residual ozone concentration at the channel exit and the logarithmic degradation factor of Bacillus subtilis spores S at the channel exit, with S defined as

$$S = -\log\left(\frac{N}{N_0}\right) \tag{21}$$

where N_0 (CFU kg^{-1} of liquid phase) is the number of Bacillus subtilis spores colony-forming unit per kg of liquid phase at the entrance of the channel. The maximum theoretical dissolved ozone concentration $C^L_{O_3,max}$ (mg kg^{-1} of liquid phase) is calculated by expressing a total ozone mass balance on chamber 1 and assuming that all the ozone initially present in the gas phase is transferred the liquid and does not decompose. This is written as

$$C^L_{O_3,\mathrm{max}} = \frac{Q^G \rho^G C^G_{O_3}}{Q^L \rho^L} \tag{23}$$

where Q^G (m^3 h^{-1}) is the gas flow rate and Q^L (m^3 h^{-1}) is the liquid flow rate.

Table 2: Impact of operating conditions on O_3 residual concentration and Bacillus subtilis spores degradation efficiency

Simulation case	Operating conditions						Maximum dissolvedO3 concentration (mg l−1)	DissolvedO3 concentration at chamber 1 outlet(mg l−1)	O3residual liquid concentration at outlet(mg. l−1)	S (-)
	Bubble diameter (mm)	Temperature (°C)	Water flow rate (m^3 h^{-1})	Gas flow rate (Nm^3 h^{-1})	Inlet gas O_3 concentration (g Nm^{-3})	kd (s^{-1})				

S1	Variable	8	2790	73	27	3.59×10^{-4}	0.70	0.61	0.55	0.66
S2	Variable	20	2900	48	20	6.03×10^{-4}	0.33	0.24	0.21	0.83
S3	Variable	20	2900	48	20	0.00×10−4	0.33	0.27	0.27	1.00
S4	Fixed	20	2900	48	20	6.03×10^{-4}	0.33	0.28	0.25	0.93
S5	Variable	20	2900	72	40	6.03×10^{-4}	0.99	0.76	0.67	2.00
S6	Variable	**5**	2900	48	20	3.13×10^{-4}	0.33	0.26	0.24	0.20
S7	Variable	5	2900	72	40	3.13×10^{-4}	0.99	0.89	0.84	0.59
S8	Variable	20	1450	48	20	6.03×10^{-4}	0.66	0.51	0.34	2.44

The S1 case corresponds to one set of operating conditions considered as usual conditions by the Vivaqua engineers.

Fig. 4a and b present simulated concentrations of ozone in the gas and in the liquid phase, respectively, in a cross section along the channel. Fig. 4a shows that the highest value of gas ozone concentration occurs at the injection point. The gas ozone concentration decreases quickly so that the gas ozone concentration is only equal to 1 g Nm^{-3} at channel midheight in chamber 1. Fig. 4a also showed that some bubbles are dragged from chamber 1 to chamber 2. In Fig. 4b, the average dissolved ozone concentration in chamber 1 is 38% lower than the concentration in chamber 2 (respectively, 0.45 and 0.60 mg l^{-1}). This is mainly due to the transport of the dissolved ozone at the base of chamber 1 directly to chamber 2. The dissolved ozone concentration decreases after chamber 2 due to ozone decay.

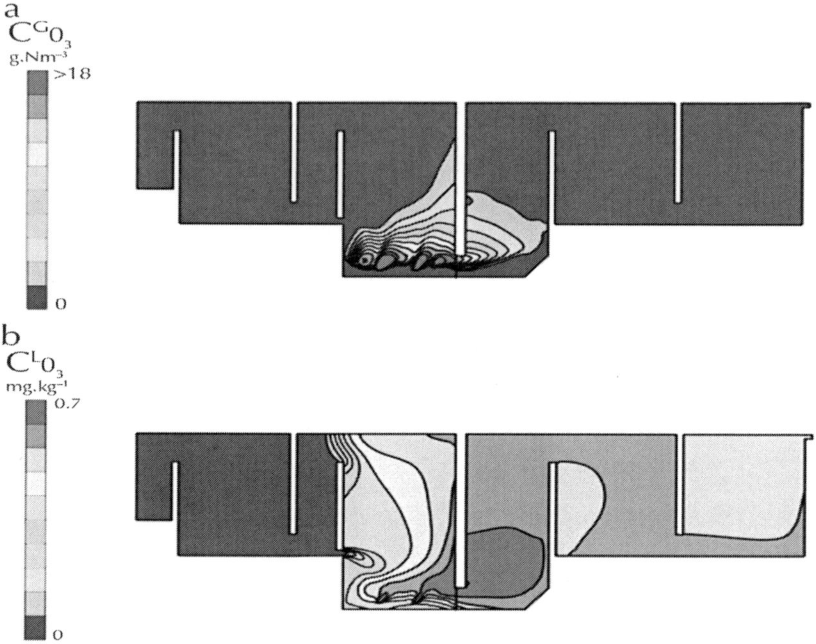

Figure 4: (a) Simulated gas ozone concentration in a cross section along the channel. ($C^G_{O_3}$ g Nm^{-3}) and (b) simulated dissolved ozone concentration in a cross section along the channel ($C^L_{O_3}$ g kg^{-1}).

The industrial value of residual ozone concentration at the channel exit is about 0.50 mg l^{-1}. The result of the simulation provides a residual concentration equal to 0.55 mg l^{-1}, which is about 10% higher than the experimental value. This tends to show that the model allows a correct evaluation of the ozone residual concentration. However, a dedicated validation should be considered to really assess the validity of the tool.

A degradation logarithm of 0.66 is predicted in this case. For this case, the low value of S highlights shows the unsuitability of current standard conditions to eliminate high concentrations of resistant microorganisms.

In order to set a reference case for the following comparisons, the operating conditions regarding ozone injection (gas flow rate and ozone concentration) are set to minimum values at 20 °C (case S2 in Table 1). In this case, the ozone residual concentration value at chamber 1 outlet is 12% higher than the concentration at the channel exit due to the ozone decomposition in the channel. The value at the chamber 1 can also be compared to the theoretical maximal ozone concentration, as defined by Eq. (22). This theoretical ozone concentration is higher than the dissolved concentration measured at chamber 1 outlet, which accounts for ozone decomposition (Table 1). A case without ozone decomposition is resolved (S3 case) and the ozone residual concentration at chamber 1 outlet is equal to 0.27 mg l^{-1}. This shows that more than 80% of the ozone is transferred from the gas to the liquid compared to the theoretical value. This confirms the efficiency of the design regarding ozone mass transfer.

Regarding degradation, this case is also unsatisfactory, as only a logarithmic degradation of 0.83 only is achieved.

For optimisation purpose, it can be concluded that enhancing mass transfer might slightly improve the treatment, but as only 20% of the ozone is not transferred with the nowadays conditions, those improvement will only be able to generate a 20% rise of the residual concentration, which is insufficient to solve the spore degradation problem.

Impact of Bubble Size Variation on Ozone Gas–liquid Mass Transfer

The S4 case corresponds to the same set of operating conditions as the S2 case. However, in the S4 case, the bubble volume is fixed.

In the S2 case, the bubble diameter which is equal to 3.3 mm at the sparger increases up to almost 3.9 mm, leading to a bubble volume increase of 60%. Table 3 presents average values of ^{G}, a and k^{L} a in chamber 1 for the S2 and S4 cases. The global volume fraction of gas in chamber 1 is 17% lower in the S4 case than in the

S2 case. This is due to the fact that the bubble volume (in the S2 case) is increasing as the bubbles are rising and that the bubble terminal velocity in both cases is similar (0.21 and 0.2 m s^{-1}, respectively, for the S2 case and the S4 case). This gap in α^G value has an impact on the interfacial area which is lower by about 12% in S4 case compared to the S2 case. As the k^L value is the same between both cases, the k^L a value also decreases by 12%. However, Table 2 shows that the residual ozone concentration is almost 20% higher in the S4 case. This is due to the mass transfer driving force, defined as $m\rho^G C^G_{O_3} - \rho^L C^L_{O_3}$, that is higher in the S4 case than in the S2 case. The mass transfer driving force is plotted against the height z_b in chamber 1 for both cases in Fig. 5. It shows that the mass transfer driving force is 23% higher in the S4 case than in the S2 case due to the dilution of ozone gas concentration in the S2 case. From Fig. 5 and Table 2, it can be seen that the higher driving force of the S4 case largely compensates the reduction of the k^L a value compared to the S2 case. Thus, the overall residual ozone concentration in case S4 is overestimated and the degradation logarithm is also slightly higher (0.93 instead of 0.83 for the S2 case).

Table 3: Mean value of α^G, a and k^L a in chamber 1

Simulation case	Gas volume fraction (%)	Interfacial area (a) (m^{-1})	($k^L a$) (s^{-1})
S2	0.23	3.92	1.46×10^{-3}
S4	0.19	3.45	1.28×10^{-3}

It is thus relevant to consider bubble size variation due to static pressure to correctly estimate the ozone gas liquid mass transfer and the residual ozone concentration.

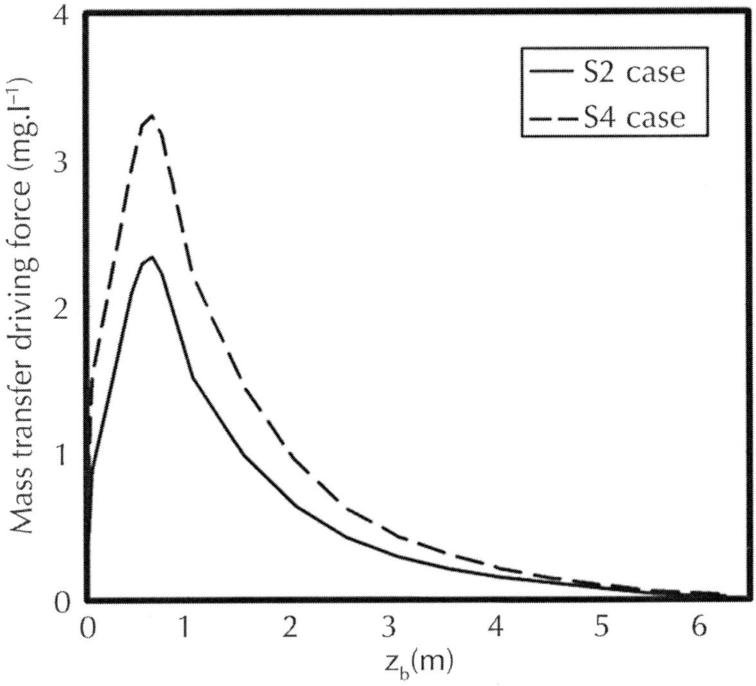

Figure 5: Variation of mass transfer driving force according to the height.

Application of the Tool to Different Operating Conditions

Four other cases, of industrial interest, have been tested to estimate the impact of a high gas flow rate and high gas concentration (maximum ozonation capability of the channel) (S5 case), the impact of a lower temperature (S6 case), an increase of both gas flow rate and inlet gas ozone concentration (S7 case) and a lower liquid flow rate (S8 case) on residual ozone concentration and degradation efficiency. All those cases are compared to the S2 case. The results are summarised in Table 2.

Maximum Ozonation Capability at 20 °C

The S5 case considers the same conditions as the S2 case but with the gas flow rate and gas ozone concentration at the gas sparger set to the maximum value as can be achieved in the Vivaqua ozonation channel. The increase of gas flow rate and gas ozone concentration leads to a residual ozone concentration three times higher than in the S2 case. As noticed in the S2 case, the ozone concentration at the chamber 1 outlet is 12% higher than the concentration at the channel exit. The dissolved ozone concentration at the chamber 1 outlet represents 77% of the maximum theoretical concentration. As far as the logarithmic degradation is concerned, its value is multiplied by two units compared to the S2 case. In this condition, the ozonation channel is able to reduce microorganism concentration by 99%.

Lower Temperature

The S6 case considers the same conditions as the S2 case but with a 5 °C temperature, a common winter temperature for water in the Vivaqua plant. The residual ozone concentration is improved by almost 15% compared to the S2 case. This can be explained by a higher ozone solubility and lower value of kinetic constant for ozone decomposition (Table 1). However, this increase of ozone concentration is not important enough to counterbalance the slowdown of Bacillus subtilis spores degradation kinetic. Indeed,Table 1 shows that the Bacillus subtilis spores degradation kinetic constant k_μ and latency constant kl are lowered, leading to a logarithmic degradation factor at this temperature that is as low as 0.2 at the ozonation channel exit. Kim (2007) also noticed a similar impact of the temperature on the degradation kinetic.

Maximum Ozonation Capability at 5 °C

The S7 case highlights the impact of increasing the gas flow rate and ozone concentration at the gas sparger as can be done in the

Vivaqua ozonation channel. These operating conditions allow multiplying the residual concentration by 3.5 compared to the S6 case. The degradation logarithm is increased by almost 0.4 units but its value still remains too low to face a peak concentration of, for instance of 3×10^5 CFU ml^{-1}, microorganism at the inlet of the ozonation channel.

For now, at low temperature, the installation seems not to be able to cope with a peak concentration of resistant microorganism in the water.

Lower Liquid Flow Rate

The S8 case is a hypothetic case in order to characterise the impact of the reduction of the HRT on residual ozone concentration and on the degradation efficiency. Reducing the liquid flow rate by a factor two leads to a 60% higher residual ozone concentration at the outlet due to a longer contact time of a liquid element with the gas compared to the S2 case. The dissolved ozone concentration at the chamber 1 outlet represents 77% of the maximum theoretical concentration. The impact of a higher HRT can be noticed on the decrease of the residual ozone concentration from chamber 1 to the outlet, which is reduced by 30% in the S8 case whereas in the case S2, where the HRT is twice smaller, the decrease of the residual ozone concentration is around 15% from chamber 1 to the outlet.

This higher dissolved ozone concentration and the higher contact time of the Bacillus subtilis spores with ozone lead to a higher value of the logarithmic degradation (S=2.44).

CONCLUSIONS AND FURTHER WORK

In this paper, a tool for assessing the degradation of microorganism in ozonation channels for drinking water production is presented.

This tool is to be used for risk analysis and optimisation analysis, particularly regarding the upcoming European legislation.

This tool is based on a model of transport equations coupled to reaction kinetic models for ozone decomposition and microorganism degradation. The model explicitly includes a microorganisms degradation kinetic including a dependency to the local ozone concentration and a latency term, explicitly depending on the contact time of the microorganisms with ozone. Also, the evolution of the bubble size with the static pressure in the channel is taken into account. To our knowledge, no other ozonation model included those behaviours.

The tool is applied to an ozonation channel of the Belgian drinking water producer Vivaqua. The tool parameters values are estimated based on literature (mass transfer coefficient, ozone solubility ratio in water and ozone diffusivity coefficient in water) or through dedicated experiments on Vivaqua plant water (decomposition of ozone and disinfection kinetics, bubbles diameter).

The tool is partially validated by comparing simulated RTDs to experimental ones. Moreover, a similar order of magnitude for residual ozone is observed between the tool prediction and an experimental value in usual operating conditions. Without being a rigorous validation of the model, those elements tend to show that the tool can be used qualitatively and semi-quantitatively for predictions.

Uses of the tool show that the bubble diameter variation has a significant impact on the ozone gas–liquid transfer rate and subsequently on the residual ozone concentration and degradation efficiency. Practically, the tool predicts that the existing Vivaqua ozonation channel can barely face upcoming European legislation regarding high concentrations of resistant microorganisms.

Indeed, a potentially acceptable degradation is achieved only when injecting the maximal quantity of ozone. Moreover, temperature has a critical effect on microorganisms degradation especially if they present a latency time towards reaction with dissolved ozone. The tool indicates that those problems have to be solved by increasing the hydraulic retention time in the channel rather than trying to enhance the mass transfer efficiency.

Although useful to predict general trends, the obtained tool is sensitive to the choice of the inactivation kinetics. Moreover, the tool assumptions should be challenged by more experimental works and tests with some other microorganisms. At the same time, focus should also be made on disinfection by-product, such as bromate formation.

ACKNOWLEDGMENTS

The authors want to acknowledge the Vivaqua for their industrial collaboration and the Brussels Capital Region for their financial support.

REFERENCES

1. Audenaert, W.T.M., Callewaert, M., Nopens, I., Cromphout, J., Vanhoucke, R.,Dumoulin, A., Dejans, P., Van Hulle, S.W.H., 2010. Full-scale modelling of anozone reactor for drinking water treatment. Chemical Engineering Journal 157,551–557.
2. Bader, H., Hoigne´, J., 1981. Determination of ozone in water by the indigo method. Water Research 15, 449–456.
3. Beltran, F.J., 2004. Ozone Reaction Kinetics for Water and Wastewater systems. Lewis publisher, CRC presses company.
4. Bartrand, T.A., Farou, B., Haas, C.N., 2009. Countercurrent gas/liquid flow andmixing: Implications for water disinfection. International Journal of MultiphaseFlow 35, 171–184.
5. Calderbank, P.H., Moo-Young, M.B., 1961. The continuous phase heat and mass transfer properties of dispersions. Chemical Engineering Science 16, 39–54.
6. Chauveheid, E., 2004. Control of halogenated by-products during surface waterpotabilisation. In: Lichtfouse, J., Schwarzbauer, D.Robert (Eds.), EnvironmentalChemistry:

Green Chemistry and Pollutants in Ecosystems. Springer-Verlag,Berlin, Germany, pp. 361–370.

7. Cockx, A., Do-Quang, Z., Line´, A., Roustan, M., 1999. Use of computational fluiddynamics for simulating hydrodynamics and mass transfer in industrialozonation towers. Chemical Engineering Science 54, 5085–5090.
8. Dehaeck, Sam, 2007. Development of Glare Point, Shadow and InterferometricPlanar Techniques for Gas Bubble Sizing. Ph.D. thesis, University of Gent,Belgium.
9. Do-Quang, Z., Cockx, A., Line´, A.,, Roustan, M., 1999. Computational fluid dynamicsapplied to water and wastewater treatment facility modelling. EnvironmentEngineering and Policy 1, 137–147.
10. European Union Council Directive 98/83/EC. 3 November 1998. The quality ofwater intended for human consumption. Official Journal of the EuropeanCommunities 330, 32–54.
11. Fayolle, Y., Cockx, A., Gillot, S., Roustan, M., He´duit, A., 2007. Oxygen transferprediction in aeration tanks using CFD. Chemical Engineering Science 62,7163–7171.
12. Fluent Inc., 2006. Fluent 6.3 User's Guide Lebanon, New Hampshire (USA).Gong, X., Takagi, S., Huang, H., Matsumoto, Y, 2007. Anumerical study of masstransfer of ozone dissolution in bubble plumes with an Euler–Lagrangemethod. Chemical Engineering Science 62, 1081–1093.
13. Higbie, R., 1935. The rate of absoprtion of a pure gas into a still liquid during shortperiods of exposure. Transactions of the American Institute of ChemicalEngineers 35, 365–389.
14. ISO 7932, 2004. Microbiology of food and animal feeding stuffs—Horizontalmethod for the enumeration of presumptive Bacillus cereus—Colony-counttechnique at 301C. ISO 7932, International Organisation for Standardisation,Geneva.
15. Johnson, P.N., Davis, R.A., 1996. Diffusivity of ozone in water. Journal of ChemicalEngineering Data 41, 1485–1487.

16. Kim, D.I., 2007. Development and application of integrated ozone contactor designand optimisation tools. PhD thesis, Georgia Institute of Technology, USA.
17. Larcy A., 2006. Contribution a l'e ´tude du transfert de matiere dans les colonnes abulles de grande taille. Postgraduate thesis, Universite´ Libre de Bruxelles,Belgium.
18. Levenspiel, O., 1999. Chemical Reaction Engineering 3rd edition John Wiley andsons, New-York.
19. Morris, J.C., 1988. The aqueous solubility of ozone—A review. Ozone news 1, 14–16.Rennecker, J.L., Marinas, B.J., Rice, E.W., Owens, J.H., 1999. Inactivation of ~Cryptosporidium parvum oocyst with ozone. Water Research 33 (11),2481–2488.
20. Talvy, S., Cockx, A., Line´, A., 2007a. Modelling hydrodynamics of gas–liquid airliftreactor. AIChE Journal 53, 316–326.
21. Talvy, S., Cockx, A., Line´, A., 2007b. Modeling of oxygen mass transfer in aGas–Liquid airlift reactor. AIChE Journal 53, 335–353.
22. Van der Helm, A.W.C, Rietveld, L.C., Baars, E.T., Smeets, P.W.M.H., van Dijk, J.C.,2008. Modelling disinfection and by product formation during the initialand the second phase of natural water ozonation in a pilot-scale plugflow. Journal of water supply research and technology—aqua 57 (6),1–15.
23. Wols, B.A., Hofman, J.A.M.H., Uijttewaal, W.S.J., Rietveld, L.C., van Dijk, J.C, 2010.Evaluation of different disinfection calculation methods using CFD. EnvironmentalModelling and Software 25, 573–582.
24. Zhang J., 2006. An integrated design approach for improving drinking water ozonedisinfection treatment based on computational fluid dynamics. PhD thesis,University of Waterloo (Canada).

Chapter 7

Industrial Products and Wastes as Adsorbents for Sulphate and Chloride Removal from Synthetic Alkaline Solution and Mine Process Water

Evgenia Iakovleva[a], Ermei Mäkilä[b], Jarno Salonen[b], Maciej Sitarz[c], and Mika Sillanpää[a]

[a]Laboratory of Green Chemistry, LUT Chemtech, Lappeenranta University of Technology, Sammonkatu 12, FI-50130 Mikkeli, Finland

[b]Laboratory of Industrial Physics, Department of Physics and Astronomy, University of Turku, FI-20014 Turku, Finland

[c]Department of Silicate Chemistry, Faculty of Materials Science and Ceramics, AGH University of Science and Technology, Al Mickiewicza 30, 30-059 Krakow, Poland

ABSTRACT

Two limestones and two solid wastes were tested as adsorbents for chloride and sulphate ion removal from synthetic and real alkaline process mining water. The composition and surface structure of the adsorbents were analysed by ED-XRF, XRD, SEM and MIR spectroscopy and the nitrogen adsorption method. The optimal adsorption time and amount of adsorbent were determined by batch analysis with an anion concentration of 20 mg/L. In synthetic solution adsorbents removed 99% of SO_4^{2-} and 96% of Cl^- and in real process water the removal rate was 85% and 74% for SO_4^{2-} and Cl^-, respectively. It was found that equilibrium could successfully be fitted to the Langmuir, Toth and BiLangmuir adsorption isotherms.

INTRODUCTION

Water is a vital factor in many industrial processes, including mining. Water is much more than a medium for cooling, creating steam, transport or cleaning; it is often an essential component of the process or product [1] and [2].

As water shortages increase, so does the need to use unconventional supplies such as seawater or wastewater for industrial processes. The water deficit may also be met by energy intensive water recovery systems such as cooling water systems for industrial processes. With the right treatment technologies, water can be reused over several cycles which helps to bridge the gap between water supply and demand [1] and [3].

Process water treatment problems can be solved using a variety of treatment technologies, such as the electro-chemical method [4], [5], [6] and [7], biological degradation [8], membrane filtration [9] and [10], coagulation [11], [12] and [13], adsorption or ion exchange [10], [14], [15], [16] and [17]. Low-cost process water recovery is one of the most significant issues facing industry today. Ion exchange is considered to be one of the best and most efficient methods for removing anions from water due to its high efficiency, simplicity and low cost.

Process water content depends on the type of industrial process. Alkaline mining process waters require special treatment mainly due to the combination of a high concentration of anions, such as Cl^-, SO_4^{2-}, NO_3^-, NO_2^-, NH_4^+, and a low concentration of metal ions [1].

The removal of anions from highly alkaline process water is expensive and quite challenging. Sulphate and chloride removal is often accomplished thorough an ion exchange mechanism with limestone [10], [18] and [19]. This process involves ion exchange between OH^- on the adsorbent surface and removed anions. Natural lime is often used for process and wastewater treatment [20], [21], [22], [23] and [24]. Sulphate removal from water by adsorption has not always been successful, for example, in A. Darbi et al.'s research on sulphate removal from drinking and groundwater by bentonite [10]. Solid waste, however, is similar in composition to limestone and has a demonstrably high capacity to adsorb anions from wastewaters, yet it has received the least amount of research attention [25] and [26]. This work took a novel approach to the removal of sulphate and chloride ions from alkaline process water. Two limestone products and two solid wastes from various industries were used as adsorbents. Two challenges were faced; real mining process water treatment forms SO_4^{2-} and Cl^-, and it is difficult to reuse solid wastes. The adsorption capacities of unmodified solid wastes were studied and the effect of both competing ions on the adsorption process was also investigated.

MATERIALS AND METHODS

Materials

Two solid wastes, iron sand (RH) and pulp and paper waste (DI-60), were used. RH and DI-60 were obtained from Ekokem Company and Metsä Tissue Corporation (Finland), respectively. Two limestone products, filter sand (FS) and flotation fines (FF), obtained from Nordkalk Corporation (Finland) were applied as adsorbents. Before use all the adsorbents were milled, washed with distilled water, and oven dried for 12 h at 80 °C. The dried adsorbents were stored in glass flasks. Sodium sulphate (Merck) and sodium chloride (Sigma–Aldrich) were used for the preparation of sulphate and chloride solutions. Solution pH adjustment was carried out with 0.1 M NaOH and HCl (Merck). Milli-Q ultrapure water was used to prepare the stock and dilute solutions.

Characterisation of Adsorbents

The particle size distribution and zeta potential of the samples were determined using dispersion analyser LUMiSizer 610/611 (LUM GmbH) and ZetaSizer Nano ZS (Malvern Instruments Ltd., UK), respectively. 1.5 g of solid sample was conditioned in 50 ml of 0.001 M NaCl solutions at various final pH for 24 h in a shaking bath. Zeta potential was calculated from electrophoretic mobility measurements using the Smoluchowski equation; each value was an average of approximately 20 measurements. The pH of the suspension was adjucted using 0.01–1.0 M HCl. All solutions were prepared using ultrapure water and solution pH was determined with a pH Electrode SenTix 81.

The physical structure of the samples was analysed with N_2 adsorption at −196 °C with TriStar 3000 (Micromeritics Inc.). The specific surface area was calculated from the adsorption isotherm using Brunauer-Emmett-Teller (BET) theory [27].

Infrared spectroscopic measurements in middle infrared (MIR) spectra of adsorbents were recorded with a Bruker Vertex 70v spectrometer. MIR: transmission technique, samples were prepared by the standard KBr pellets method. Spectra were collected in the mid infrared region (4000–400 cm^{-1}) averaging 124 scans with 4 cm^{-1} resolution. Samples were prepared according to the standard KBr pellets method.

The microstructure of the materials were examined using scanning electron microscopes (SEM, Nova Nano SEM 200, FEI Company and Hitachi S-4800), with energy dispersive X-ray spectroscopy (EDX, EDAX) for micro-area analysis. The experiment was carried out at low vacuum conditions in secondary electron mode and samples were covered with a layer of gold.

X-ray diffraction (XRD) data were collected with a PANalytical Empyrean powder diffractometer using Cu Kα radiation. The configuration was a standard Bragg–Brentano (θ/θ) reflection setup with a Ni-filter placed in front of the detector. The measurements were performed in a θ scanning range of 3–90° with a 0.007° step size and 69 s of measurement time for each step at ambient temperature and pressure.

The chemical composition of materials was determined by an energy dispersive X-ray fluorescent (ED XRF) analyser X-Art (Joint Stock Company Comita, St. Petersburg, Russia) [28].

Batch Adsorption Experiments

Ion exchange tests were conducted by mixing a known weight of FF, FS, RH and DI-60 with 15 ml of synthetic sulphate and chloride ions in alkaline solution. The ion concentrations of the synthetic solution ranged from 20 to 2000 mg/l. The solutions were shaken in a mechanical shaker ST5 (CAT M.Zipper GmbH, Staufen, Germany) from 1 to 720 h. 2 ml samples of the solution were taken from flasks at known time intervals and then filtered using a 0.20 μm diameter polypropylene syringe filter. All tests were performed at ambient temperature (22 ± 2 °C) and pressure. The final concentrations of sulphate and chloride anions in the solutions were determined with high-performance liquid chromatography (HPLC). A Shimadzu HPLC equipped with a conductivity detector (Shimadzu Model CDD-10A) was used (column: 4.0 mm ID × 250 mmL Shodex IC SI-50 4E; mobile phase: solution of 3.2 mM Na_2CO_3 and 1 mM $NAHCO_3$in ultrapure water; flow rate: 0.7 ml/min; temperature:ambient). Adsorption was calculated as:

$$A = \frac{C_i - C_t}{C_i} * 100, \qquad (1)$$

where A is adsorption (%), C_i and C_t (mg/l) are the ion concentrations in the initial and post treatment solutions, respectively.

Sorbent Optimisation

The optimum amount of sorbent was determined by mixing 15 ml of SO_4^{2-} and Cl^- synthetic solutions with different amounts of sorbent (2, 5, 10, 20, 30, 40, and 50 g/l). Each result is the average of three samples.

Contact Time Optimisation

The optimum contact time was determined by mixing 40 g/l of FF, FS DI-60, and RH in 15 ml flasks tumbled at 100 rpm from 1 to 720 h with an initial concentration of SO_4^{2-} and Cl^- of 20 mg/L. Samples were taken after 1, 2, 4, 8, 24, 48, 96, 120, 240, 480, and 720 h.

RESULTS AND DISCUSSION

Adsorbent Characterisation

The chemical compositions of the studied materials are shown in Table 1. These results were obtained by ED-XRF and XRD. Each material contains calcium (12–37%). RH contains more than 17% sulphur. Such a high sulphur concentration could affect the removal of sulphate ions, because sulphur on the adsorbent surface can initiate competition in the ion exchange process [13]. All chemical compounds structurally similar to calcite should have high adsorptive properties for sulphate and chloride ions.

Table 1: Chemical composition of adsorbents (XRF and XRD analysis)

Element (%)	FS	FF	RH	DI-60
Ca	21.3	37.2	14.4	12.0
Fe	1.07	0.43	7.2	0.46
K	0.254	0.241	0.3	0.329
Mn	0.0216	0.0100	0.27	0.0043
S	0.100	0.105	17.6	0.039
Si	42	10.3	0.2	4.4
Sr	0.026	0.074	0.021	0.016
Ti	0.052	0.042	2.3	0.059
Rb	0.0013	0.0005	0.001	0.0020
Compounds formula	$CaMgSi_2O_6$	$(K,H_3O)Al_2Si_3AlO_{10}(OH)_2$	$Ca(SO_4)(H_2O)$	$Ca(CO_3)$
	$CaCO_3$	$CaSiO_3$	$Ca(CO_3)$	$Al_2(Si_2O_5)(OH)_4$
	$(K,H_3O)Al_2Si_3AlO_{10}(OH)_2$			
Compounds name	Diopside	Illite	Calcium sulphate hydrate	Calcite
	Calcite	Parawollastonite	Calcite	Kaolinite

The particle size distribution and specific surface area of the original materials are shown in Table 2. The particle size of FS and FF adsorbents are approximately the same (approximately 500 μm). The particle size for DI-60 is about three times larger (≈1500 μm), and for RH it is four times smaller (≈125 μm). Unexpectedly, the active surface area of RH is significantly larger than that of other adsorbents, which may be explained by its relatively high porosity. The results of spectroscopic analysis of the adsorbent are shown in Fig. 1. Each image presents two spectra, before (1) and after (2) washing and drying the adsorbents. As can be seen, pretreatment does not change the distribution and location of functional groups on the adsorbent surface. Spectroscopic studies in the mid-infrared (MIR) region confirm the results of the XRD studies. The spectra of each adsorbent show two additional bands characteristic of water molecules, at about 1620–1640 and 3450–3480 cm^{-1} (Fig. 1). The OH^- stretch vibrations of the SiOH group adsorb at 3200–3700 cm^-, and strong Si–O bands at 830–1100 cm^- are observed in the spectra of DI-60 (Fig. 1c). The spectra of sulphones show strong adsorption bands at 1120–1160 cm^- for RH (Fig. 1d). These bands arise from symmetric SO_2[29]. Adsorbent surfaces were studied by XRD (Table 1) and MRI (Table 2). Both methods showed the presence of sulphur on the RH surface and silicone on the FF, FS, and DI-60 surfaces.

Table 2: Characteristic of the surface of adsorbents

Name of adsorbent	Particle size (μm)	Specific surface area (m^2/g)	Zeta-potential (mV)
			pH = 11
FF	599	1.35	0
FS	451	1.82	−25
RH	180	62.5	−7
DI-60	1390	3.29	−14

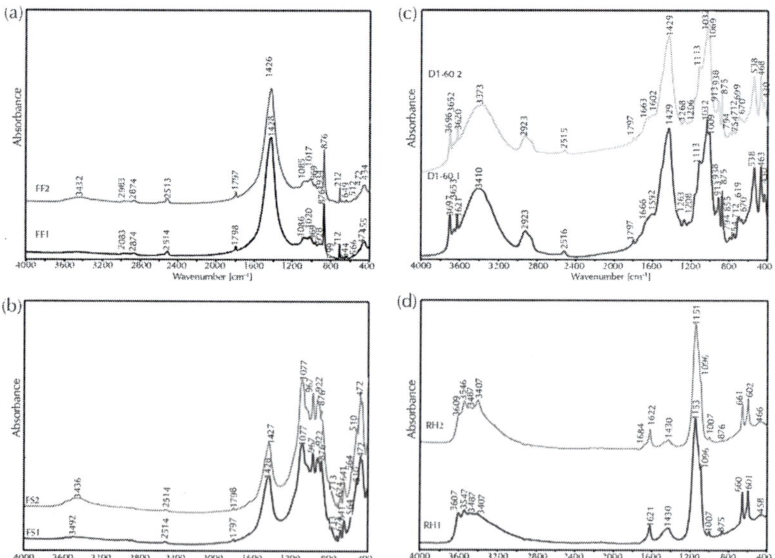

Figure 1: MID spectrums of FF (a), FS (b), DI-60 (c) and RH (d).

SEM images are presented in Fig. 2. The structure of RH is the finest, which is partly due to its highly specific surface area. This is confirmed by data on the active surface of adsorbents obtained by the BET method (Table 2); RH has the largest active surface area.

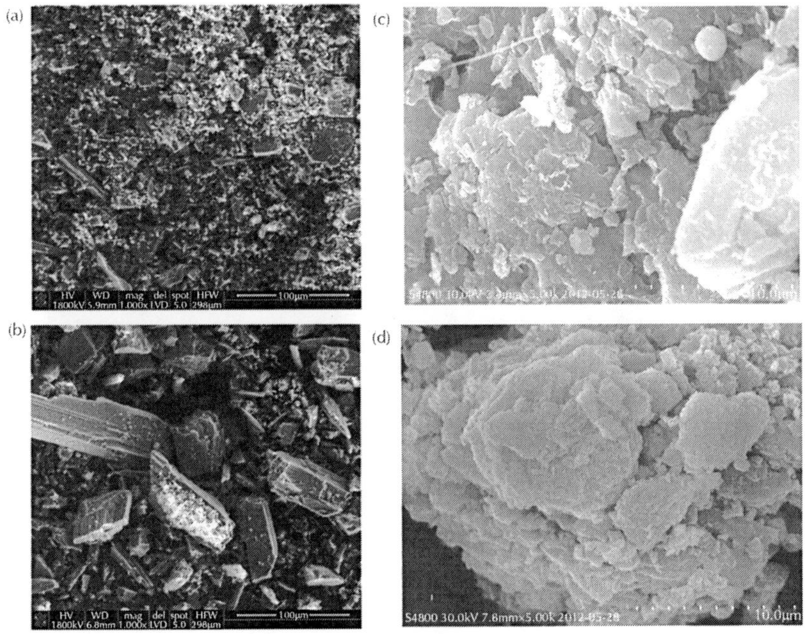

Figure 2: SEM images of FF (a), FS (b), DI-60 (c) and RH (d).

Optimisation of Adsorbent Amount and Ion Competition

The optimal amount of adsorbent for chloride and sulphate removal with an initial concentration of 20 mg/l is shown in Fig. 3. This indicates that initially Cl^- and SO_4^{2-} removal increased with increasing amounts of adsorbent, with equilibrium reached at an adsorbent concentration of 40 g/l. The highest removal percentage of sulphate (99%) was achieved with DI-60; sulphate removal was 88.5%, 90%, and 92% with RH, FS, and FF, respectively. Kolics et al. described the competition between sulphates and chlorides during the adsorption on aluminium [30]. They found that chlorides were adsorbed better than sulphates. Our experiments show that in case of adsorption with DI-60 and RH, ion competition was not observed: Cl^-ion removal reached approximately 96% with DI-60 and RH. With limestone-like adsorbents (FF and FS), however, this was not the case. Chloride removal was substantially lower than that of sulphates; only 66% and 54% were removed with FF and FS, respectively. Chloride was adsorbed better than sulphate with RH, which had the highest adsorption capacity. This phenomenon can consider as one of the factor influencing for better adsorption of Cl^- ions than SO_4^{2-} ions, because chloride ions are smaller than sulphate [12]. Since RH has a porous surface structure it was better able to adsorb chloride.

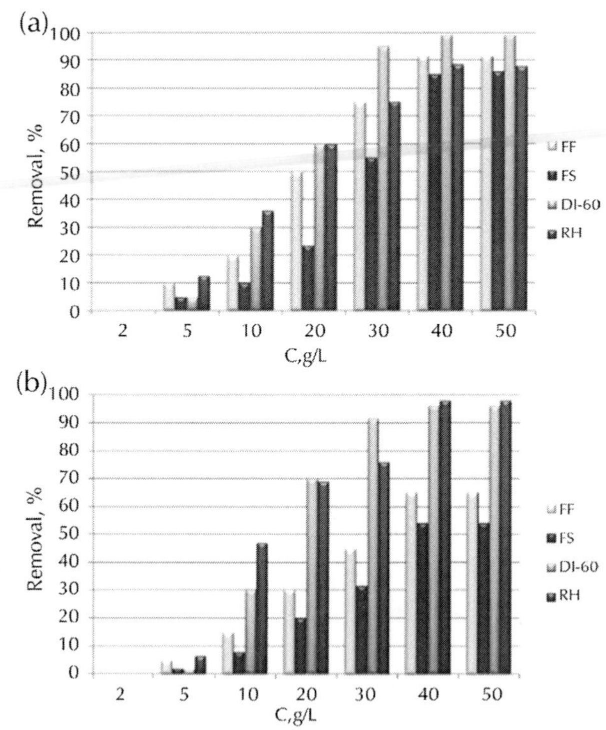

Figure 3: Effect of dosage of FF, FS, DI-60 and RH for removal of SO_4^{2-} (a) and Cl^- (b) with initial concentration of ions 20 mg/l.

Contact Time Optimisation

It was found that the removal percentage increases with time, reaching the maximum at 480 h (Fig. 4). The maximum amount of Cl^- and SO_4^{2-} adsorbed to FF, FS, DI-60, and RH was 65% and 92%; 90% and 54%; 96% and 99%; 95% and 89%, respectively. It was observed that the ion exchange process does not change the initial pH of the solution [12] and [13].

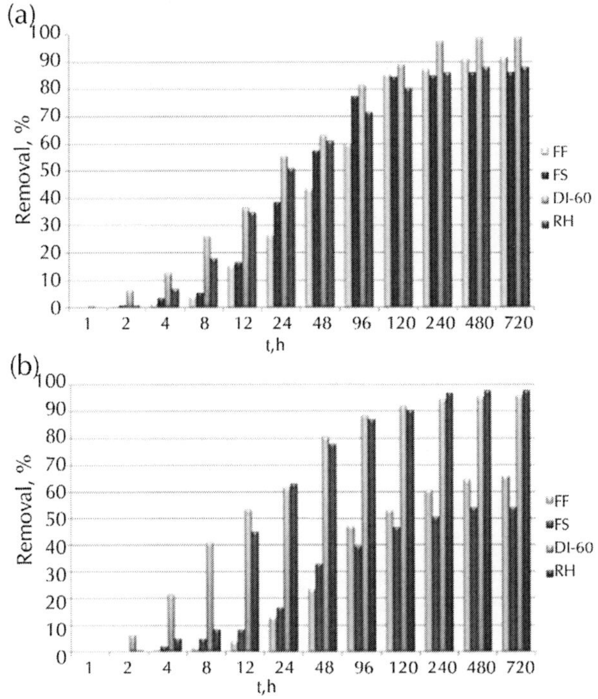

Figure 4: Kinetics of SO_4^{2-} (a) and Cl^- (b) ions removal with FF, FS, DI-60 and RH, at pH 12. Initial ions concentration is 20 mg/L.

Adsorption Isotherms

Equilibrium of the adsorbate sorption on the adsorbent surface at a constant temperature is described with empirical models known as adsorption isotherms. Mathematical modelling of adsorption isotherms gives us information about the adsorption mechanism. A rigorous theory of adsorption isotherms was proposed by Langmuir for monolayer adsorption on specific homogeneous sites without attractive forces between the adsorbate molecules and their mobility along the surface [31]. Further development of the theory eliminates assumptions made by Langmuir. Since the adsorbents have a complex structure, homogeneous as well as heterogeneous adsorption mechanisms are possible. Two-parameter Langmuir isotherm (Eq. (2)) was used to describe a classic case of adsorption in a homogeneous system. Three-parameter (Toth Eq. (3)) and four-parameter (BiLangmuir Eq. (4)) models were selected to describe adsorption process in the case of a heterogeneous system:

$$q_e = \frac{q_m K_L C_e}{1 + K_L C_e} \tag{2}$$

$$q_e = \frac{q_m C_e}{(a_T + C_e^{m_T})^{\frac{1}{m_T}}} \tag{3}$$

$$q_e = \frac{q_{m1} K_{L1} C_e}{1 + K_{L1} C_e} + \frac{q_{m2} K_{L2} C_e}{1 + K_{L2} C_e} \tag{4}$$

where q_e and C_e are anion-exchange capacity (mmol/g) and concentration of anions in the solution (mmol/L) at equilibrium, respectively; q_m is the maximum uptake capacity; K_L is Langmuir, K_{L1} and $K_{L2}$are the BiLangmuir constants related to the energy of anion-exchange; a_T is the adsorptive potential constant (mmol/L) and m_T the heterogeneity factor of the Toth isotherm.

Isotherm parameters were calculated with minimisation of the error distribution between experimental and predicted data by Marquardt's percent standard deviation (MPSD):

$$\sum_{i=1}^{n}\left(\frac{q_{e,exp}-q_{e,calc}}{q_{e,exp}}\right)_i^2 \tag{5}$$

The Langmuir isotherm is the most frequently used theory [32]. A basic assumption is that one molecule occupies a site and no further adsorption can take place at that site. In the other words, at equilibrium, saturation point is reached where no further adsorption can occur. Parameter q_m shows the number of adsorbent sites that actively interact with anions of the sorbate. The *K* coefficient is the affinity between sorbate and adsorbent. According to Langmuir theory the affinity between the studied adsorbents and adsorbates is:

$$\text{RH; FS; DI-60}:\ SO_4^{2-} > Cl^-$$

$$\text{FF}:\ Cl^- \geqslant SO_4^{2-}$$

The number of sites (q_m) which participate in the adsorption of Cl^- is much greater than in the adsorption of SO_4^{2-}. This could be because chloride molecules are smaller than sulphate molecules; Langmuir theory implies that the amount of adsorbed chloride will be much greater [33]. Yet since the affinity (*K*) of adsorbents to sulphate ions is higher, better removal of sulphate could be observed with FF, FS and DI-60. Chloride was more effectively removed with RH, since its pores are significantly larger than those of other sorbents (Table 2). The combination of large RH pores and small chloride ions allows for near complete chloride removal (Fig. 3b). Comparison with published data was difficult, because very few studies of the concurrent removal of chloride and sulphate ions exist; however, Hendricks found the same relationship between sulphate and chloride ions [34].

The Toth isotherm (three-parameter equation) is an adaptation of Langmuir theory to allow for a heterogeneous system. The Toth equation is obtained by adding two parameters, (a_T) and (m_T), to the Langmuir equation [35]. Those two parameters allow for heterogeneity of system. The Toth isotherm fits adsorption of both anions with DI-60 slightly better than the Langmuir one (Fig. 5), apparently because DI-60 is more heterogeneous.

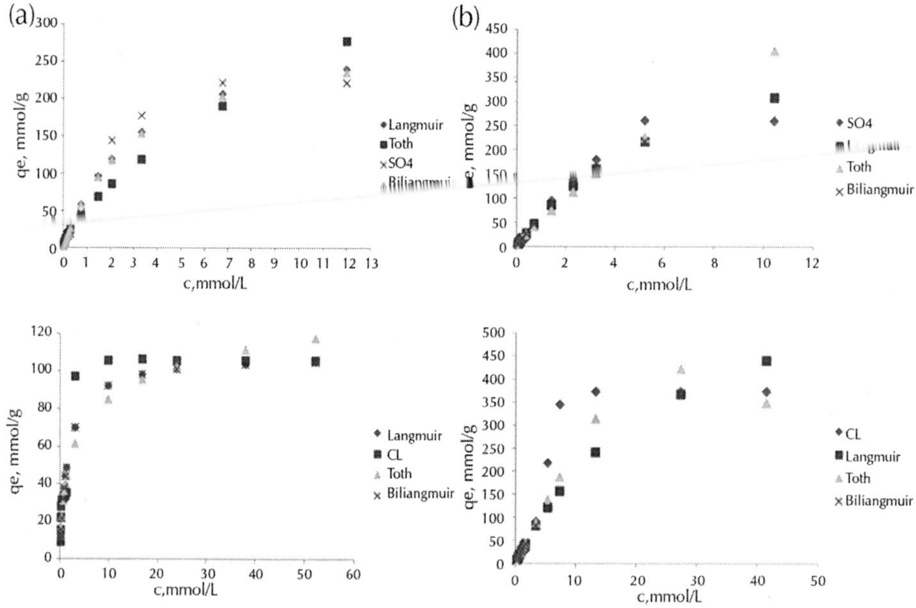

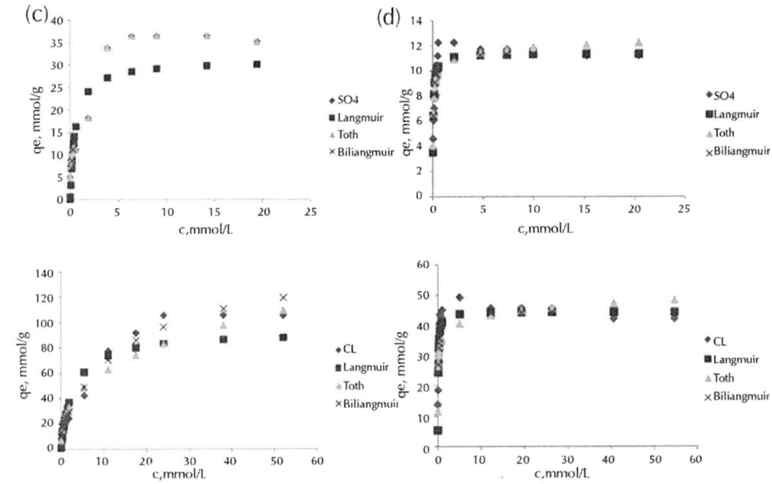

Figure 5: Adsorption isotherms for SO_4^{2-} and Cl^- with FF (a), FS (b), DI-60 (c) and RH (d) (40 g/L). Contact time is 480 h, initial ions concentration is 20 mg/L.

The BiLangmuir isotherm assumes that the surface contains two divergent active sites having different affinities towards the target compound [36]. BiLangmuir isotherms for both anions fit better to the FF, FS, and DI-60 adsorbents (Fig. 5), possibly because these adsorbents have at least two anion adsorption sites. The BiLangmuir model was rather efficient in reflecting the adsorption process.

Anion Removal from Real Process Water

Chloride and sulphate ion concentrations in real mining process water were 2.5 and 30 g/L, respectively. The chemical composition of the process waster is presented in Table 3. The removal of anions was conducted by batch method at ambient temperature for 480 h with 40 g/L of each adsorbent. The experiment was carried out in three steps, using fresh adsorbent at each stage, and the total process took 60 days. Chloride removal with FF, FS, DI-60, and RH was 52%, 45%, 80%, and 82%, respectively; sulphate removal with the same adsorbents was 73%, 68%, 54%, and 75%, respectively (Fig. 6). Increasing adsorbent concentration and time did not increase anion removal, as might be expected. Equilibrium occurred at the optimal adsorption time (480 h) and adsorbent concentration (40 g/L). The lower removal of anions from the real process water could be due to its complex composition. It contains other ions beside Cl^- and SO_4^{2} which can act as competitors during the adsorption process.

Table 3: Isotherm parameters for FS, FF, DI-60 and RH during adsorption process of anions from synthetic solution

		C_e(mmol/L)	**$q_{e\,exp}$(mmol/g)**	**$q_{e\,model}$(mmol/g)**	**q_m (mmol/g)**		**K_L (L/mmol)**		**R^2**
Langmuir									
FF	SO_4^{2-}	6.77	221	215	285		0.453		0.998
	Cl^-	16.9	106	104	115		0.583		0.984
FS	SO_4^{2-}	5.2	261	217	526		0.135		0.998
	Cl^-	13.2	373	241	718		0.038		0.997
DI-60	SO_4^{2-}	8.96	29.13	36.46	30.85		1.895		0.928
	Cl^-	23.94	105	83	93		0.355		0.928
RH	SO_4^{2-}	4.74	11.72	11.24	11.37		18.68		0.728
	Cl^-	5.07	49.3	43.8	44.5		12.06		0.757
		C_e(mmol/L)	$q_{e\,exp}$(mmol/g)	$q_{e\,model}$ (mmol/g)	K (L/mmol)		a_T(mmol/L)	m_T	R^2
Toth									
FF	SO_4^{2-}	6.8	221.4	190.2	10,936		0.019	0.338	0.995
	Cl^-	16.9	106.3	102.2	141.5		0.821	0.854	0.997
FS	SO_4^{2-}	5.2	260.7	225	114.5		0.207	0.177	0.999
	Cl^-	13.2	373.3	314.1	35.8		10,149	2.67	0.997
DI-60	SO_4^{2-}	6.4	36.5	25.9	29.2		14.3	0.941	0.985
	Cl^-	23.9	105.6	82.8	60,389		0.001	0.641	0.981
RH	SO_4^{2-}	0.55	12.2	9.9	616		0.035	0.957	0.897
	Cl^-	5.07	49.3	40.7	3810		0.019	0.929	0.961
		C_e(mmol/L)	$q_{e\,exp}$ (mmol/g)	$q_{e\,model}$ (mmol/g)	q_{m1} mmol/g	q_{m2} (mmol/g)	K_1 (L/mmol)	K_2(L/ mmol)	R^2
BiLangmuir									
FF	SO_4^{2-}	6.8	221.4	218	118.5	166.4	0.162	0.101	0.995
	Cl^-	16.9	106.3	98.2	95	32.9	0.133	16.7	0.989
FS	SO_4^{2-}	5.2	260.7	217.2	139.8	123.5	0.089	0.033	0.998
	Cl^-	13.2	373.3	241.2	271.4	120.5	0.038	0.213	0.997

DI-60	SO_4^{2-}	6.4	36.5	28.5	37.7	8.03	0.257	78.9	0.991
	Cl^-	23.9	105.6	96.4	17.6	132.1	140.9	0.154	0.991
RH	SO_4^{2-}	0.55	12.2	10.4	11.0	0.106	20.7	0.107	0.775
	Cl^-	5.07	49.3	43.8	3.7	42.3	1.7	14.1	0.825

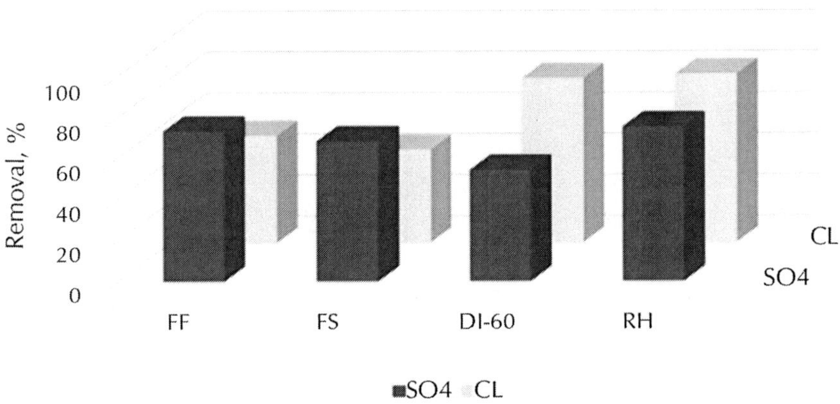

Figure 6: Removal of SO_4^{2-} and Cl^- with FF, FS, DI-60 and RH from real process water. Initial concentration of SO_4^{2-} and Cl^- was 30 and 2.5 g/L, respectively. Amount of adsorbents is 40 g/L.

CONCLUSIONS

The efficiency of four solid wastes used as anion exchangers to remove chloride and sulphate ions from alkaline mining process water was investigated. The optimal adsorption parameters were estimated with synthetic solution. The optimal time (480 h) and adsorbent concentration (40 g/L) were determined by batch method. The maximum adsorption capacities of FF, FS, DI-60 and RH ranged from 221, 261, 29 and 12 mmol/g for SO_4^{2-} removal, to 106, 373, 105 and 49 mmol/g for Cl^- removal, respectively. All three isotherms (Langmuir; Toth; BiLangmuir) described both anion exchange processes well. The FF and FS removed more than 50% of both anions and RH removed approximately 80% of sulphate ions. The DI-60 removed more than 80% of both sulphate and chloride ions. All adsorbents could be used industrially for alkaline process water treatment: although the removal process is long (around 2 months), the treatment requires a minimum amount of energy and is low in cost due to the use of solid wastes as adsorbents.

ACKNOWLEDGMENTS

The authors acknowledge the financial support of the Finnish Funding Agency for Technology and Innovation (TEKES). The authors would also like to thank Nordkalk, Norilsk Nickel Harjavalta and Ekokem for the materials and Norilsk Nickel for the process water. The authors are grateful to Dr. V. Vergizova for her thoughtful and constructive review of this paper. Eduard Musin is acknowledged for his contribution in performing experiments.

REFERENCES

1. D. Kemp, C.J. Bond, D.M. Franks, C. Cote, Mining, water and human rights: making the connection, J. Cleaner Prod. 18 (15) (2010) 1553–1562.
2. P.M. Heikkinen, P. Noras, R. Salminen, U.-M. Mroueh, P. Vahanne, M.M. Wahlström, T. Kaartinen, M. Juvankoski, E. Vestola, E. Mäkelä, T. Leino, M. Kosonen, T. Hatakka, J. Jarva, T. Kauppila, J. Leveinen, P. Lintinen, P. Suomela, H. Pöyry, P. Vallius, J. Nevalainen, P. Tolla, V. Komppa, Environmental Techniques for the Extractive Industries, Mine Closure Handbook, Espoo, 2008.
3. J.M. Anderson, Current water recycling initiatives in Australia: scenarios for the 21th century, Water Sci. Technol. 33 (10–11) (1996) 37–43.
4. X. Wu, Zh. Liu, X. Liu, Chloride ion removal from zinc sulfate aqueous solution by electrochemical method, Hydrometallurgy 134–135 (2013) 62–65.

5. J.-H. Yeo, J.-H. Choi, Enhancement of nitrate removal from a solution of mixed nitrate, chloride and sulfate ions using a nitrate-selective carbon electrode, Desalination 320 (2013) 10–16.

6. J.M. Paz-Garcia, B. Johannesson, L.M. Ottosen, A.B. Ribeiro, J.M. RodriguesMaroto, Simulation-based analysis of the differences in the removal rate of chlorides, nitrates and sulfates by electrokinetic desalination treatments, Electrochim. Acta 89 (2013) 436–444.

7. R.S. Gärtner, F.G. Wilhelm, G.J. Witkamp, M. Wessling, Regeneration of mixed solvent by electrodialysis: selective removal of chloride and sulfate, J. Membrane Sci. 250 (1–2) (2005) 113–133.

8. M.D. Krom, A.B. David, E.D. Ingall, L.G. Benning, S. Clerici, S. Bottrell, C. Davies, N.J. Potts, R.J.G. Mortimer, J. van Rijn, Bacterially mediated removal of phosphorus and cycling of nitrate and sulfate in the waste stream of a "zerodischarge" recirculating mariculture system, Water Res. 56 (2014) 109–121.

9. R. Malaisamy, A. Talla-Nwafo, K.L. Jones, Polyelectrolyte modification of nanofiltration membrane for selective removal of monovalent anions, Sep. Purif. Technol. 77 (3) (2011) 367–374.

10. A. Darbi, T. Viraraghavan, Y.-G. Jin, L. Braul, D. Corkal, Sulfate removal from water, Water Qual. Res. J. Canada 38 (2003) 169–182.

11. B. Gao, W. Zhou, Q. Yue, L. Lui, Y. Wang, Al-Ferron kinetics and quantitative calculation of Al(III) species in polyaluminum chloride coagulation, Colloids Surf B 278 (2006) 235–240.

12. [12] D. Tang, H. Wang, J. Gregory, Relative importance of charge neutralization and precipitation on coagulation of Kaolin with PAC: effect of sulphate ion, Environ. Sci. Technol. 36 (2002) 1815–1820.

13. R. Silva, L. Cadorin, J. Rubio, Sulphate ions removal from an aqueous solution: I. Co-precipitation with hydrolysed

aluminum-bearing salts, Miner. Eng. 23 (2010) 1220–1226.

14. JJ. Dron, A. Dodi Comparison of adsorption equilibrium models for the study of Cl^-, NO_3^- and SO_4^- removal from aqueous solution by an anion exchange resin J. Hazard. Mater., 190 (1–3) (2011), pp. 300–307
15. R. Haghsheno, A. Mohebbi, H. Hashemipour, A. Srrafi, Study of kinetic and fixed bed operation of removal of sulphate anions from an industrial wastewater by an anion exchange resin, J. Hazard. Mater. 166 (2–3) (2009) 961–966.
16. P.-L. Sang, Y.-Y. Wang, L.-Y. Zhang, L.-Y. Chai, H.-Y. Wang, Effective adsorption of sulphate ions with poly(m-phenylenediamine) in aqueous solution and its adsorption mechanism, Trans. Nonferrous Met. Soc. China 23 (1) (2013) 243– 252.
17. A. Halajnia, S. Oustan, N. Najafi, A.R. Khataee, A. Lakzian, Adsorption– desorption characteristics of nitrate, phosphate and sulphate on Mg–Al layered double hydroxide, Appl. Clay Sci. 80–81 (2013) 305–312.
18. J. Rubio, R.D.R. Silva, A.N. Silveira, Treatment of acid mine drainage (AMD) in South Brazil. Comparative active process and water reuse, Int. J. Miner. Process. 93 (2009) 103–109.
19. H.Z. Zhao, C. Liu, Y. Xu, J.R. Ni, High-concentration polyaluminium chloride: preparation and effects of the Al concentration on the distribution and transformation of Al species, Chem. Eng. J. 155 (2009) 528–533.
20. T.C. Zhang, D.G. Lampe, Sulfur: limestone autotrophic denitrification process for treatment of nitrate-contaminated water: batch experiments, Water Res. 33 (1999) 599–608.
21. A.M. Silva, E.C. Cunha, F.D.R. Silva, V.A. Leao, Treatment of high-manganese mine water with limestone and sodium carbonate, J. Clean. Prod. 29–30 (2013) 11–19.
22. A.M. Silva, R.M.F. Lima, V.A. Leao, Mine water treatment with limestone for sulfate removal, J. Hazard. Mater. 221–222 (2012) 45–55.

23. A. Sdiri, T. Higashi, F. Jamoussi, S. Bouaziz, Effect of impurities on the removal of heavy metals by natural limestones in aqueous systems, J. Environ. Manage. 93 (2012) 245–253.
24. S. Hussain, H.A. Aziz, M.H. Isa, M.N. Adlan, F.A.H. Assari, Physico-chemical method for ammonia removal from synthetic wastewater using limestone and GAC in batch and column studies, Bioresour. Technol. 98 (2007) 874–880.
25. M.A. Rahman, Sh. Ahsan, S. Kaneco, H. Katsumata, T. Suzuki, K. Ohta, Wastewater treatment with multilayer media of waste and natural indigenous materials, J. Environ. Manage. 74 (2005) 107–110.
26. A. Bhatnagar, M. Sillanpää, A review of emerging adsorbents for nitrate removal from waste, Chem. Eng. J. 168 (2011) 493–504.
27. S. Brunauer, P. Emmett, E. Teller, Adsorption of gases in multimolecular layers, J. Am. Chem. Soc. 60 (1938) 309–319.
28. A.S. Serebryakov, E.L. Demchenko, V.I. Koudryashov, A.D. Sokolov, Energy dispersive X-ray fluorescent (ED XRF) analyser X-Art for investigation of artworks, Nucl. Instrum. Methods Phys. Res. B 213 (2004) 699–702.
29. Silverstein, Spectroscopy Chemistry NMR FTIR MS.
30. A. Kolics, J.C. Polkinghorne, A. Wieckowski, Adsorption of sulphate and chloride ions on aluminum, Electrochem. Acta 43 (18) (1998) 2605–2618.
31. A. Dąbrowski, Adsorption – from theory to practice, Adv. Colloid Interface Sci. 93 (2001) 135–224.
32. S.J. Allen, G. Mckay, J.F. Porter, Adsorption isotherm models for basic dye adsorption by peat in single and binary component systems, J. Colloid Interface Sci. 280 (2004) 322–333.
33. L. Lv, P. Sun, Zh. Gu, H. Du, X. Pang, X. Tao, R. Xu, L. Xu, Removal of chloride ion from aqueous solution by ZnAl-NO3 layered double hydroxides as anionexchanger, J. Hazard. Mater. 161 (2009) 1444–1449.
34. D. Hendricks, Water Treatment Unit Process, Taylor and Francis Group, Boca Ration, 2006.

35. A.P. Terzyk, J. Chatlas, P.A. Gauden, G. Rychlicki, P. Kowalczyk, Developing the solution analogue of the Toth adsorption isotherm equation, J. Colloid Interface Sci. 266 (2003) 473–476.

36. E. Repo, M. Mäkinen, S. Rengaraj, G. Natarajan, A. Bhatnagar, M. Sillanpää, Lepidocrocite and its heat-treated forms as effective arsenic adsorbents in aqueous medium, Chem. Eng. J. 180 (2012) 159–169.

Chapter 8

Control of a Coagulation Chemical Dosing Unit for Water Treatment Plants using MMPC based on Fuzzy Weighting

Oladipupo Bello, Yskandar Hamam, and Karim Djouani

Electrical Engineering Department/F'SATI, Tshwane University of Technology, Pretoria X680, South Africa

ABSTRACT

A multiple model predictive control (MMPC) scheme based on fuzzy weighting is proposed to control the coagulation chemical dosing unit for water treatment plants. The system is represented by a set of linearised models, where each model corresponds to one of the operating regions of the dosing unit. MPC local controller is designed for each local model. The local control input sequence of each model/controller pair in the scheme is aggregated by fuzzy weighting into a global control input to regulate the surface charge and pH level of the dosing unit. The parameters of the fuzzifiers are fine tuned to improve the performance of the proposed control scheme using the nonlinear curve fitting optimisation technique. Simulation results on a process model of the coagulation chemical dosing unit show the ability of the proposed fuzzy weighting MMPC algorithm to transit smoothly from one operating region to another effectively better than switching and weighting MMPC schemes.

INTRODUCTION

In water treatment process, it is often desirable to optimise the coagulants and pH adjusting chemicals added to raw waters during the coagulation process by regulating the flow rates of these chemicals. The merits of adequate coagulation control include reduction in the wastage of expensive chemicals and operational cost, and realisation of portable water quality targets. Coagulation control therefore plays a vital role in water treatment plant operations. The process has a challenging control problem because it is nonlinear and complex.

Several studies have been carried out to apply different control strategies to the coagulation process [1], [8] and [14]. In Adgar et al. [1], the authors investigated the application of feedback control on coagulation process in a twin pilot plant using streaming current detector and pH sensor to improve the existing control strategy. The authors proposed a new decoupling control scheme that reduced the interactions between the pH and coagulant dosage loops.

The new PI control strategy was found to be less susceptible to disturbance when compared with the separate feedback control loops. In [8], a fuzzy logic based control scheme was developed to realise the optimum dosing rate control according to variations in the raw water quality characteristics. The control scheme was designed using self-learning algorithm based on Takagi and Sugeno's fuzzy inference rule and was programmed in a small-scale programmable logic controller.

In another study carried out by Paz and Ocampo [14], the authors compared the performance of PID controller with linear model predictive control (MPC) on a single-input, single-output model of a coagulant dosage system. However, the study does not consider the effect of pH on the coagulation control. In addition, when MPC is applied to a nonlinear and multivariable plant over the entire range of operating conditions, its performance deteriorates. This is due to the difficulty regarding the development of a single mathematical model to describe the behaviour of the system in its entire range of operating conditions.

One of the methods to address this problem is to describe the entire system by a set of models instead of using a single model. Each local model in the set is valid in a narrow region around one of the operating points. MPC local controller can be designed for each local model and aggregated to form a global nonlinear control strategy using a classic switching or intelligent technique [11], [15], [13], [6], [7] and [21].

In this paper, applications and performance evaluation of multiple model predictive control (MMPC) strategies for the regulation of surface charge (SC) and pH of a coagulation chemical dosing unit for water treatment plants is investigated. The process model of the chemical dosing unit at the Rietvlei water treatment plant in the City of Tshwane, South Africa is developed for computer simulation tests of the MMPC schemes.

The control objective is to maintain the output variables at the desired values by manipulating the flow rates of the primary coagulant, co-coagulant and pH adjustment chemicals. The proposed control scheme in this paper combines the concepts of

MMPC and fuzzy logic to control the dosing unit over a range of operating conditions. Models of the coagulation chemical dosing unit at three operating conditions or regions (underdose, normal and overdose) are developed based on the characteristics of the water flowing out of the dosing unit. The proposed control strategy is compared with the classic switching and weighting MMPC schemes under different scenarios to evaluate its performances.

The next section presents the coagulation chemical dosing unit at the Rietvlei water treatment plant and its process model. Section 3 describes the concept of multiple model predictive control scheme and the proposed MMPC based on fuzzy weighting. The simulation results and related discussions are presented in Section 4. The concluding remarks are stated in the last section.

MATERIALS AND METHOD

Description of the Coagulation Chemical Dosing Unit

Fig. 1 illustrates the coagulation chemical dosing unit of Rietvlei water treatment plant. It is made up of a concrete mixing tank with inlet and outlet channels. There are four metering or dosing pumps that are used for dosing chemicals into the influent raw waters flowing into the treatment plant. Two of the pumps are on active position and the remaining two pumps are on redundant position. One of the active pumps feeds polyelectrolyte solution while the other feeds ferric chloride solution into the concrete mixing tank. The polyelectrolyte solution is referred to as sudfloc 3835, a blend of epichlorohydrin/dimethylamine (polyamine) and aluminium chlorohydrate [12]. The maximum flow rates of pump are specified as 0.0075 L/s. The pumps are controlled by programmable logic controllers (PLCs) that are connected to the plant's supervisory control and data acquisition (SCADA) system. The quantities of the coagulation chemicals dosed depend on the flow rate of the raw water flowing into the plant. Calcium hydroxide (hydrated lime) in slurry form is also added to the mixing tank using a diaphragm pump to stabilise the water and adjust the pH value to the desired level.

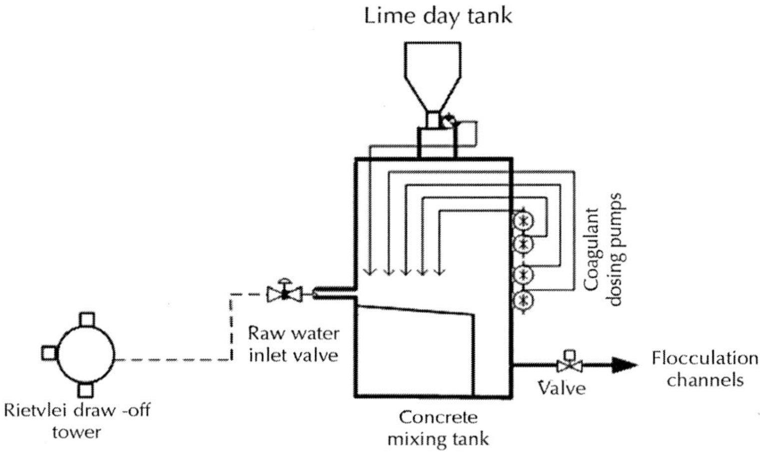

Figure 1: Coagulation chemical dosing unit of Rietvlei water treatment plant.

Nonlinear Model of the Chemical Dosing Unit

The coagulation chemical dosing unit involves a nonlinear process. The dynamics of the process model are expressed as follows [2]:

$$2(C_5H_{12}ONCl)_n + 2Al_2Cl(OH)_5 + 2Ca(HCO_3)_2 \longrightarrow 2(C_5H_{12}ON^+)_n + 4Al(OH)_3 + OH^- + 2CaCl_2 + 4CO_2 \tag{1}$$

$$2FeCl + 2Ca(HCO_3)_2 \longrightarrow 2Fe(OH)_3 + 3CaCl_2 + 6CO_2 \tag{2}$$

$$Ca(HCO_3)_2 + Ca(OH)_2 \longrightarrow 2CaCO_3 + 2H_2O \quad (3)$$

Chemical reactions in (1), (2) and (3) take place simultaneously. Therefore, the overall chemical equation becomes:

$$2(C_5H_{12}ONCl)_n + 2Al_2Cl(OH)_5 + 2FeCl_3 + Ca(OH)_2 + 6Ca(HCO_3)_2$$
$$\longrightarrow 2(C_5H_{12}ON^+)_n + 2OH^- + 2Fe(OH)_3 + 4Al(OH)_3 + 5CaCl_2$$

$$+ 10CO_2 + 2CaCO_3 + 2H_2O \quad (4)$$

The reaction invariants of (4) are $[n_{(C5H12ON+)}]$, $[Al^{3+}]$, $[Fe^{3+}]$, $[Ca^{2+}]$, $[H^+]$, $[HCO^-_3]$, $[OH^-]$,$[SO^{2-}_4]$, $[Cl^-]$ and $[CO^{2-}_3]$. However, $[Cl^-]$, $[SO^{2-}_4]$ and $[CO^{2-}_3]$ ions present in the system do not take part in the neutralisation reactions [2]. Therefore, the electroneutrality equation of (4) is expressed as:

$$[(C_5H_{12}ON^+)_n] + [Al^{3+}] + [Fe^{3+}] + [Ca^{2+}] + [H^+] = [HCO_3^-] + [OH^-] \quad (5)$$

Rewriting (5) gives:

$$[HCO_3^-] - [(C_5H_{12}ON^+)_n] - Al^{3+} - [Fe^{3+}] - Ca^{2+} = [H^+] - [OH^-] \quad (6)$$

Let the difference of the ionic concentrations be expressed as:

$$X = [HCO_3^-] - [(C_5H_{12}ON^+)_n + Al^{3+}] - [Fe^{3+}] - [Ca^{2+}] \quad (7)$$

Where

$$X = [H^+] - [OH^-] \tag{8}$$

Assuming that there is perfect mixing in the tank reactor, the relationship between the ionic concentrations of the effluent and input concentrations using the material balance equations can be expressed as:

$$V\frac{dC_1}{dt} = C_{1_{in}} q_a - C_1 q_{out} \tag{9}$$

$$V\frac{dC_2}{dt} = C_{2_{in}} q_b - C_2 q_{out} \tag{10}$$

$$V\frac{dC_3}{dt} = C_{3_{in}} q_c - C_3 q_{out} \tag{11}$$

$$V\frac{dC_4}{dt} = C_{4_{in}} q_{in} - C_4 q_{out} \tag{12}$$

where C_1 represents concentration of the sudfloc 3835 ions, $[C_5H_{12}ON_n^+ + Al^{3+}]$ at the mixing tank outlet, C_{1in} sudfloc ionic concentration at the mixing tank inlet, C_2 concentration of ferric ions, $[Fe^{3+}]$ at the mixing tank outlet, C_{2in} ferric ionic concentration at the mixing tank inlet, C_3 concentration of calcium ions (Ca^{2+}) at the mixing tank outlet, C_{3in} calcium ionic concentration at the mixing tank inlet,C_4 concentration of bicarbonate ions $[HCO^-_3]$

at the mixing tank outlet, C_{4in} bicarbonate ionic concentration of influent stream (raw water). q_a denotes flow rate of sudfloc 3835 solution, q_b flow rate of ferric chloride solution, q_c flow rate of hydrated lime, q_{out} flow rate of the effluent stream (chemically dosed water), q_{in} flow rate of the influent stream (raw water), V volume of mixing tank, $[H^+]$ hydrogen ions concentration, $[OH^-]$ hydroxide ions concentration and kw dissociation constant of water.

Adding (9) to (11) and then subtract the sum from (12) gives:

$$V\frac{dC_4}{dt} - V\frac{dC_1}{dt} - V\frac{dC_2}{dt} - V\frac{C_3}{dt} = C_{4_{in}}q_{in} - C_{1_{in}}q_a - C_{2_{in}}q_b - C_{3_{in}}q_c + C_1 q_{out} + C_2 q_{out} + C_3 q_{out} - C_4 q_{out} \quad (13)$$

The material balance expression of the mixing tank reactor yields:

$$V\frac{dX}{dt} = C_{4_{in}}q_{in} - C_{1_{in}}q_a - C_{2_{in}}q_b - C_{3_{in}}q_c - (C_4 - C_1 - C_2 - C_3)q_{out} \quad (14)$$

Thus, (14) can be expressed as:

$$V\frac{dX}{dt} = C_{4_{in}}q_{in} - C_{1_{in}}q_a - C_{2_{in}}q_b - C_{3_{in}}q_c - Xq_{out} \quad (15)$$

where $X = C_4 - C_1 - C_2 - C_3$.

The dissociation equation for water is

$$[H^+][OH^-] = k_w = 10^{-14} \quad (16)$$

Substituting (8) into (16) gives a quadratic expression:

$$[H^+]^2 - X[H^+] - k_w = 0 \tag{17}$$

The solutions of (17) are written as:

$$\begin{aligned}
&\text{if} X < 0, \text{then} [H^+] = \frac{X}{2}\left(\sqrt{1+\frac{4k_w}{X^2}}+1\right)\\
&\text{if} X = 0, \text{then} [H^+] = \sqrt{k_w}\\
&\text{if} X > 0, \text{then} [H^+] = -\frac{X}{2}\left(\sqrt{1+\frac{4k_w}{X^2}}-1\right)
\end{aligned} \tag{18}$$

The concentration of the hydrogen ions [H⁺] may be expressed in the logarithmic function as:

$$pH = -\log[H^+] \tag{19}$$

For the surface charge modelling of the raw water treatment, the theory of ion adsorption and distribution between the cations from the coagulant solutions and the negatively charged colloids described by Gouy–Chapman diffuse double layer theory is applied. The fundamental differential equation for the double layer is obtained as [4]:

$$\frac{d^2\psi}{dx^2} = \frac{8\pi n e^2 z^2}{\epsilon \kappa T(\sinh((zF\psi)/(RT))((zF)/(RT)))} \tag{20}$$

where e is electron charge, n number of ion pairs, ϵ relative dielectric permittivity, Boltzmann constant, electrical potential,

T absolute temperature, R universal gas constant, z valence and F Faraday constant.

However, the total surface charge, is obtained using the Poisson equation. By integrating it with respect to x and taking 0 and ∞ as the boundary conditions for x, yields:

$$\sigma = -\int_0^{\infty} \rho dx = \frac{\epsilon}{4\pi} \int_0^{\infty} \left(\frac{d^2\psi}{dx^2} \right) dx \tag{21}$$

$$\sigma = \left[\frac{2n\epsilon\kappa T}{\pi} \right]^{(1/2)} \sinh \left(\frac{zF\psi_0}{2RT} \right) \tag{22}$$

Eq. (22) shows a direct relationship between total surface charge and surface electrical potential. Using Nernst equation, the surface electrical potential is approximated as [4]:

$$\psi_0 = \left(\frac{RT}{F} \right) \ln \frac{C}{C_0} \tag{23}$$

where C is concentration of potential determining ion in solution C_0 concentration of potential determining ion at $_0$ = 0. In the coagulation process, H^+ and OH^- are regarded as potential determining ions (PDIs), thus (23) can be written as:

$$\psi_0 = \left(\frac{RT}{F} \right) \ln \frac{H^+}{H_0^+} \tag{24}$$

where H^+_0 is the point of zero charge.

By substituting Eq. (24) into (22) becomes:

$$\sigma = \left[\left(\frac{2}{\pi}\right) n\epsilon\kappa T\right]^{(1/2)} \sinh 1.15(pH_0 - pH) \tag{25}$$

where surface charge (SC), Boltzmann constant, T temperature, ϵ relative dielectric permittivity, pH_0pH at point of zero charge and n ionic strength.

Eqs. (15), (19) and (25) form a system of differential and algebraic equations (DAE) that describe the dynamics of the coagulation chemical dosing unit.

The controlled variables or measured output vector are:

$$\mathbf{y} = [SC \quad pH]^T \tag{26}$$

The manipulated variables or control input vector are:

$$\mathbf{u} = [q_a \quad q_b \quad q_c]^T \tag{27}$$

For a multiple-input, multiple-output nonlinear process, the ith local model is thus represented as:

$$\begin{aligned} \mathbf{x}_i(k+1) &= A_i\mathbf{x}_i(k) + B_i\mathbf{u}_i(k) \\ \mathbf{y}_i(k) &= C_i\mathbf{x}_i(k) + D_i\mathbf{u}_i(k) \end{aligned} \tag{28}$$

where $x \in Rn$ is n dimensional state vector, $u \in Rm$ is m dimensional manipulated vector, $y \in R^l$ is ldimensional output vector.

MULTIPLE MODEL PREDICTIVE CONTROL (MMPC) SCHEME

MMPC Structure

The multiple model predictive control (MMPC) is an effective strategy developed to control nonlinear process where a linear MPC performance may be inadequate. In addition, it is very suitable for systems or processes where the operating conditions change abruptly from one point to another. MMPC is realised by defining models that describe the different operating conditions of the system. It includes design of MPC local controller for each of the predefined models, identification of the appropriate model and selection of the appropriate control signals at each instant [10].

In general terms, MMPC consists of three parts. The first part is the model/controller bank. These are the collections of models and controllers for each of the operating conditions of the system. The second part is the supervisory or decision making unit. It uses either switching or weighting technique to decide the appropriate control input signal to the plant at a specified period of time. The last part is the design of local controllers using the model predictive control. Based on the supervisory mechanism, MMPC can thus be broadly classified into switching and weighting control techniques [19].

Switching Multiple Model Predictive Control (SMMPC)

This technique involves selection of the best controller, based on the rules of supervisor in each cycle of operation or the model with the smallest error with respect to the process. In this method, the output of the system is divided into various regions. A model and controller is developed for each region. When the system is operating at a particular operation region, the best model/controller pair for calculating control signal and controlling the process is selected.When the operating region changes, an appropriate model/controller pair would be chosen.

Fig. 2 shows the block diagram of a switching MMPC. In Fig. 2, M#1, M#2,...,M#N are the N models, which are connected in parallel with the actual process. y(k) is the output vector of the process, $u_1(k)$, $u_2(k)$, ...,$u_N(k)$ are the control actions to each of the corresponding models and the u(k) is the final control action. r(k) is the reference trajectory, $\mathbf{y}_m(k)$ is the predictive output vector of the selected model, e(k) is the error between plant output vector and predictive output vector of the selected model. SMMPC is obtained by the defining some performance indices, where the switching takes place when the performance index reaches its minimum value. The performance index is expressed as:

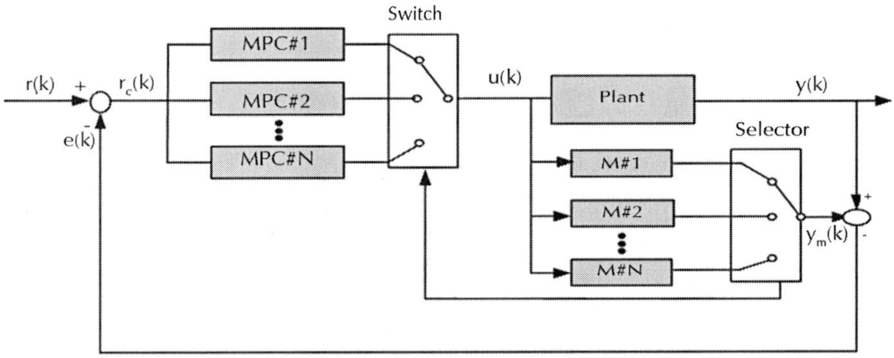

Figure 2: Switching multiple model predictive control.

$$J_m(k) = \alpha e_m^2(k) + \beta \int_0^k e^{-\lambda(k-\tau)} e_m^2(\tau) d\tau \tag{29}$$

Where

$$\begin{aligned} &\alpha \geq 0 \\ &\beta, \lambda > 0 \\ &e_m(k) = \mathbf{y}(k) - \mathbf{y}_m(k), \quad m = 1, \ldots, N \end{aligned} \tag{30}$$

The design parameters for the switching of the controller are , and , where and weighting factors on the instantaneous measures and the long term accuracy, respectively. Also is a forgetting factor which assures the boundedness of the criterion for the bounded $e_m(k)$. Selection of these parameters are vital for providing quick response during changing from one model to the other and stability of the switching scheme [11], [10] and [19].

Weighting Multiple Model Predictive Control (WMMPC)

In this approach, MPC local controllers are designed based on model representing each operating region. The final control input signal u(k), is realised based on the control strategy scheme shown in Fig. 3. It is expressed as the linear combination of the weighted outputs of the MPC local controllers:

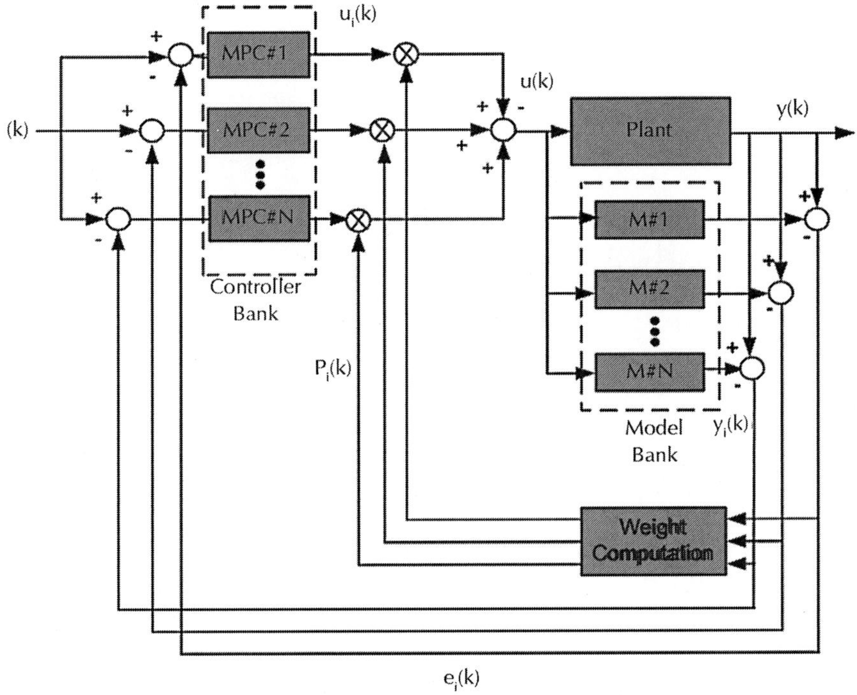

Figure 3: Weighting multiple model predictive control.

$$\mathbf{u}(k) = \sum_{i=1}^{N} P_i(k)\mathbf{u}_i(k) \tag{31}$$

Where

$$\sum_{i=1}^{N} P_i(k) = 1 \tag{32}$$

and N is the number of model/controller pairs. $P_i(k)$ is the appropriate weight of the ith MPC local controller, $u_i(k)$ is the ith MPC local controller output and u(k) is the final control output [20],[10] and [19].

Fuzzy Weighting Multiple Models Predictive Control (FWMMPC)

A weighting MMPC based on fuzzy logic concept is proposed in this paper as shown in Fig. 4. In this control strategy, each output of the MPC local controller is provided with a fuzzifier to convert the MPC output signals into fuzzy representations. These fuzzified MPC outputs are aggregated and converted back into crisp variable(s) using defuzzification method. The output variables of the defuzzifier are applied to plant as control input signals or manipulated variables. The fuzzifier in this study is made up of a product of two sigmoid membership function.The expression for the product of two sigmoid membership function is [17]:

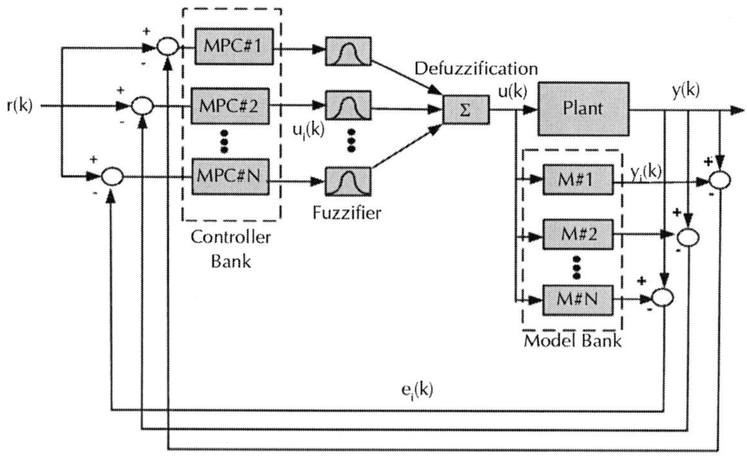

Figure 4: Fuzzy weighting multiple model predictive control.

$$f(x; a_1, c_1, a_2, c_2) = f_1(x; a_1, c_1) \cdot f_2(x; a_2, c_2) \tag{33}$$

where $f_k(x) = 1/(1 + e^{(-ak(x-ck))})$, $k = 1, 2$. k_1 and k_2 are the first and second sigmoid curves, a_1 and a_2 are the slope control parameters for the curves, c_1 and c_2 are the inflection points control for the curves.

The values of $f_k(x)$ lies between 0 and 1. The choice of this membership function in this study above other functions is based on its ability to achieve smoothness and specify asymmetric functions [17]. Furthermore, optimum performance can be obtained from these fuzzifiers by tuning the parameters of their membership functions. In this study, the nonlinear least squares curve fitting optimisation technique is used to obtain the optimal parameters of the fuzzifiers as briefly discussed.

Consider a fuzzy curve or nonlinear function:

$$y = f(x; a, b, c, d) \tag{34}$$

where a, b, c and d are the unknown parameters.

Using (34) and a set of measurement pairs (x_1, y_1), (x_2, y_2), ..., (x_n, y_n), the unknown parameters could be estimated. The procedure involves minimization of the sum of the squares of the residuals:

$$J = \sum_{1}^{n} e_i^2 = e_1^2 + e_2^2 + \ldots + e_n^2 \tag{35}$$

where $e_i = y_i - f(x_i\,; a, b, c, d)$.

Taking the partial derivatives of J with respect to each of the unknown parameters and equated them to zero gives the minimisation of the performance index, J. With further analysis, m normal equations given as follows can be obtained:

$$\sum_{i=1}^{n} e_i = 0 \tag{36}$$

$$\sum_{i=1}^{n} x_i e_i = 0 \tag{37}$$

$$\sum_{i=1}^{n} x_i^m e_i = 0 \tag{38}$$

where m is the order of the nonlinear function. From the above normal equations, the unknown parameters that will yield the optimal or most probable values can be determined [18] and [16].

Defuzzification method is implemented using the weighted average of all the fuzzifiers' output. The global controller output vector, u(t), is expressed as in [21]:

$$\mathbf{u}(k) = \frac{\sum_{i=1}^{N} \mu_i \mathbf{u}_i}{\sum_{i=1}^{N} \mu_i} \tag{39}$$

where μ_i and u_i are the membership function and fuzzified output the local MPC of the ith operating region respectively, and N is the number of the MPC local controllers.

Design and Implementation

The design and implementation of the proposed FWMMPC scheme for the dosing unit is based on the integration of MPC local controllers, fuzzifiers and defuzzifier with the local models defined in Table 1.For each linearised model, MPC local controller is designed using the procedures presented in this subsection.

Table 1: Operating regions of the dosing unit

Operating region	Surface charge (μeq/mg)	pH values (pH unit)
Underdose	$SC < 0$	$pH > 8$
Normal	$0 \leq SC \leq 1 \times 10^{-5}$	$7 \leq pH \leq 8$
Overdose	$SC > 1 \times 10^{-5}$	$pH < 7$

The control strategy of the MPC is to find a sequence of control signals $\Delta u(k)$, $\Delta u(k + 1)$, …, $\Delta u(k + M + 1)$ that yield the predicted outputs $y(k + 1|k), y(k + 2|k), \ldots, y(k + P|k)$, which is based on the model of the chemical dosing unit response, to some desired reference signal vector $r(k + 1|k), r(k + 2|k), \ldots, r(k + P|k)$. The span of the control sequence, M, is referred to as the control horizon, while that of the predicted output,P, is called the prediction horizon. Although the computed control values are based on P time steps, only the first move is implemented and a new sequence of control inputs is computed at the next sampling instant. The following cost function is chosen to implement the proposed control strategy:

$$J(u) = \underbrace{\min}_{\Delta u(k),\ldots,\Delta u(k+M-1)} \quad S_y(k) + S_u(k) + S_{\Delta u}(k) \tag{40}$$

Where

$$S_y(k) = \sum_{i=1}^{P}\sum_{j=1}^{n_y}\{w_j^y[r_j(k+i) - y_j(k+1)]\}^2 \tag{41}$$

$$S_u(k) = \sum_{i=1}^{M}\sum_{j=1}^{n_u}\{w_j^u[u_j(k+i-1) - \bar{u}_j]\}^2 \tag{42}$$

$$S_{\Delta u}(k) = \sum_{i=1}^{M}\sum_{j=1}^{n_u}\{w_j^{\Delta u}\Delta u(k+i-1)\}^2 \tag{43}$$

Subject to:

$$u_{\min} \leqslant u(k+i-1) \leqslant u_{\max}, \quad \text{for } i = 1 \text{ to } M \tag{44}$$

$$\Delta u_{\min} \leqslant \Delta u(k+i-1) - \Delta u(k+i-2) \leqslant \Delta u_{\max}, \quad \text{for } i = 1 \text{ to } M \tag{45}$$

$$y_{\min} \leqslant y(k+i) \leqslant y_{\max}, \quad \text{for } i = 1 \text{ to } P \tag{46}$$

Where

$$S_y(k) = \sum_{i=1}^{P}\sum_{j=1}^{n_y}\{w_j^y[r_j(k+i) - y_j(k+1)]\}^2 \tag{47}$$

$$S_u(k) = \sum_{i=1}^{M}\sum_{j=1}^{n_u} w_j^u[u_j(k+i-1) - \overline{u}_j]^2 \tag{48}$$

$$S_{\Delta u}(k) = \sum_{i=1}^{M}\sum_{j=1}^{n_u}\{w_j^{\Delta u}\Delta u(k+i-1)\}^2 \tag{49}$$

where r(k) is the reference vector, y(k + j|k) is the j-step ahead predicted output given the present output measurements, w^y_j is the positive definite output weighting matrix, $w^{\Delta u}_j$ is the positive semi definite input increment weighting matrix, w^u_j is the positive semi definite input weighting matrix. $\overline{u}_j$ is the nominal value of input j. The weighting matrices and the prediction horizon parameters, P, and the control horizon M are the tuning parameters which can be used to shape the closed-loop response of the system [20], [5] and [3].

The design of the MPC local controllers was meant to regulate the two outputs of the dosing unit. Equal weight was specified for the output variables. w^y_j matrix was set as:

$$= \begin{bmatrix} 1 & 0 \\ 0 & 1 \end{bmatrix} \quad (50)$$

The prediction horizon, P was chosen as 15 and the control horizon, M was chosen as 5 for all MPC local controllers. A sampling time of 1s was specified for the controllers design. The nonzero $w^u{}_j$ and $w^{\Delta u}{}_j$ matrices were used to shape the closed loop response of the unit. By penalising changes in the manipulated variables, a slower but smoother servo trajectory could be obtained. Their values were selected and adjusted during the simulation to tune the controller. These matrices were set as:

$$= \begin{bmatrix} 0.2 & 0 & 0 \\ 0 & 0.2 & 0 \\ 0 & 0 & 0.2 \end{bmatrix} \quad (51)$$

Each fuzzifier takes its input from the MPC local controller output. The fuzzifiers were defined using three linguistic variables with their associated membership functions: low; medium; and high. From several experimental trials, the best result was obtained when the underdose operating region MPC local controller was combined with the high fuzzifier, the normal operating region MPC local controller with the mediumfuzzifier and the overdose operating region MPC local controller with the low fuzzifier. The initial parameters of each fuzzifier were selected using the MATLAB® Fuzzy Logic Toolbox. The parameters of the fuzzifiers were selected such that the low fuzzifier was skewed to the left, high fuzzifier was skewed to the right and medium fuzzifier was symmetric (Table 3).

Table 2: Parameters of the fuzzifiers

Fuzzifier	Slope (a_1)	Inflection point (c_1)	Slope (a_2)	Inflection point (c_2)
Low (selected)	13.73	−0.20	−13.73	0.20

Medium (selected)	13.73	0.30	−13.73	0.70
High (selected)	13.73	0.80	−13.73	1.20
Low (initial guess)	7.50	0.10	8.50	0.30
Medium (initial guess)	10.73	0.7	9.73	1.80
High (initial guess)	9.60	1.0	8.90	1.50
Low (optimal)	−12.57	0.20	20.10	−1.08
Medium (optimal)	0.005	−293.90	10.06	0.28
High (optimal)	0.99	−2.60	13.77	0.80

However, to estimate the optimal values of the slopes and inflection points of the fuzzifiers, the MATLAB®Optimisation Toolbox was used for this purpose. A set of input vector with 100 data point was generated from a normal distribution in the interval of [0, 1]. The associated measurements (output vector) were obtained through the general expression of the product of sigmoid function in (33). Using the least squares curve fitting optimisation procedures in Section 3.4, the optimal parameters of the fuzzifiers were computed from a set of initial guess of the parameters. The selected, guessed and optimal parameters of fuzzifiers are shown in Table 2. Fig. 5 shows the three fuzzifiers used for the FWMMPC scheme.

Table 3: Process modelling parameters

Variable and symbols	**Values and units**
Sudfloc 3835 ions concentration, $([(C_5H_{12}ON^+)_n] + [Al^{3+}])$	0.00001 mol/L
Ferric ions concentration, $[Fe^{3+}]$	0.00001 mol/L
Calcium ions concentration, $[Ca^{2+}]$	0.00001 mol/L
Bicarbonate ion concentration, $[HCO^-_3]$	0.00001 mol/L
Hydrogen ion concentration, $[H^+]$	10^{-7}
Tank volume, V	29,000 L
Dissociation constant of water, k_w	10^{-14}
Temperature, T	298 K
Ionic strength, n	50×10^6 mol/L
Relative dielectric permittivity, ϵ	80
Boltzman constant, κ	1.38×10^{23} J K^{-1}

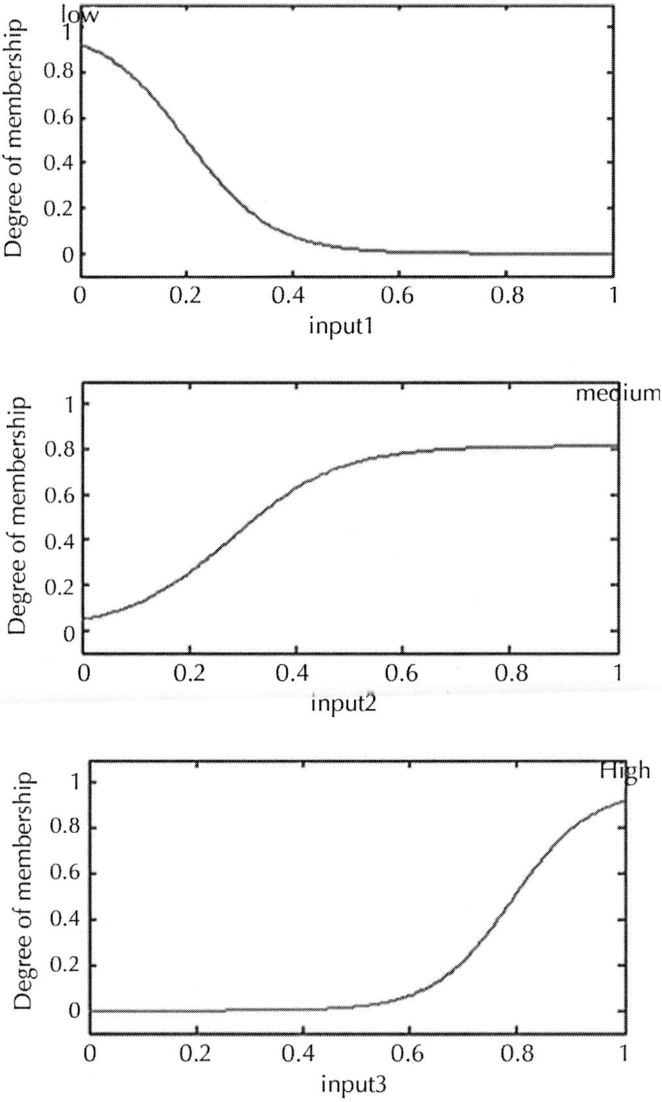

Figure 5: Product of two sigmoid functions of the low, medium and high fuzzifiers.

SIMULATION RESULTS AND DISCUSSIONS

Simulation Setup

The performance of the proposed control strategy is compared with SMMPC and WMMPC schemes using the MATLAB®/SIMULINK® version 7.10 software. For the purpose of comparison of simulation tests, all the schemes have the same MPC local controllers and operating regions but different modes of transiting from one region to another as described in Section 3.

All the MPC local controllers have the same tuning parameters. The MMPC schemes were designed to control the surface charge and pH values by manipulating the flow rates of the primary coagulant, co-coagulant and hydrated lime. In all the simulation tests discussed in the next subsection, the performance of the proposed FMMPC, SMMPC and WMMPC schemes were evaluated on the process model of the dosing unit. The performance measures of the schemes are evaluated using root mean squared error (RMSE). This is a useful index to compare the individual performance of a model with other models [9] and [21]:

$$\text{RMSE} = \sqrt{\frac{\sum_{t=1}^{T}(y_p(t) - y_s(t))}{T}} \tag{52}$$

where y_p is the controlled output vector of the dosing unit and y_s is the set point trajectory, T is the simulation period.

Simulation Results

In the first scenario, computer simulation test was performed without constraints on the variables of MPC local controllers. Fig. 6 shows the performance of the coagulation chemical dosing unit based on the proposed control (FWMMPC), SMMPC and WMMPC over its entire range of operating conditions or regions. Fig. 7 indicates the control input moves of the three MMPC schemes evaluated in this study. It can be observed from Fig. 6 that the response of the FMMPC is better than SMMPC and WMMPC schemes. In addition, the normalised values of the root mean square error for the SMMPC, WMMPC and FWMMPC are presented in Table 4 in a non-dimensional form. The RMSE values of FWMMPC are the smallest indicating its superior capability to achieve the objective of reducing control system error.

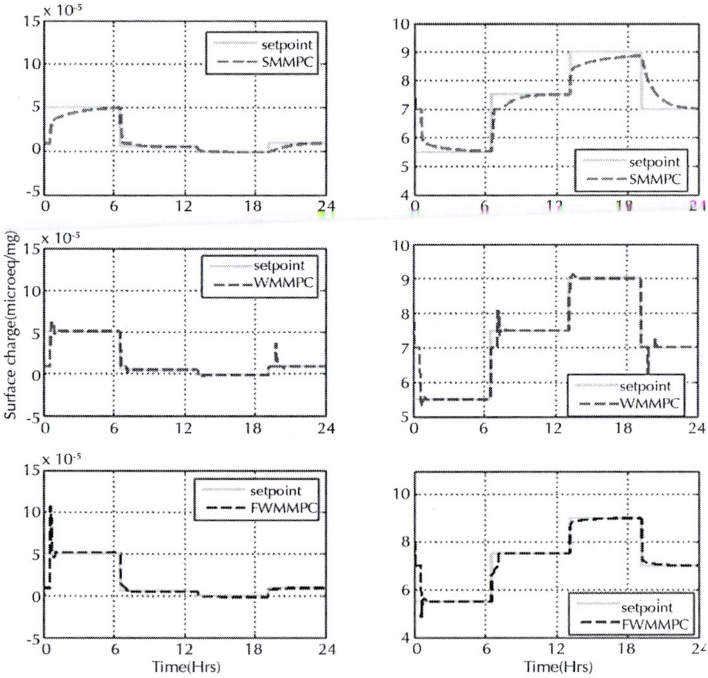

Figure 6: Performance of MMPC schemes with unconstrained MPC local controllers.

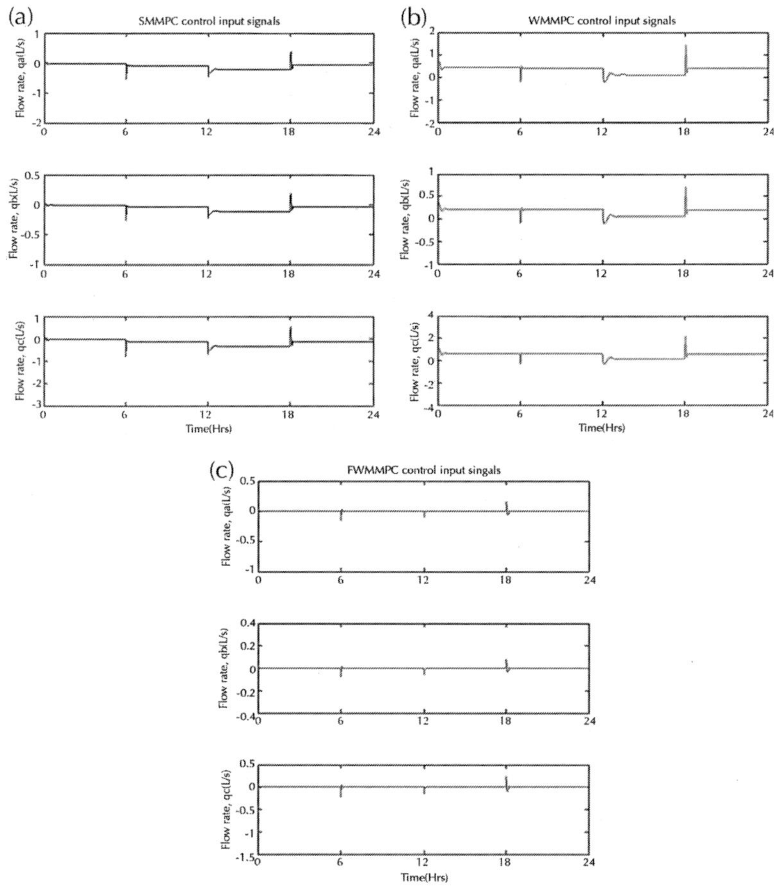

Figure 7: Control input moves of the MMPC schemes with unconstrained MPC local controllers.

Table 4: Root mean squared errors (RMSE)

Control scheme	SC_{Uncon}	pH_{Uncon}	SC_{con}	pH_{con}	SC_{Disrej}	pH_{Disrej}
SMMPC	0.089	0.112	0.101	0.211	0.095	0.105
WMMPC	0.075	0.066	0.096	0.209	0.099	0.106
FWMMPC	0.067	0.061	0.0763	0.097	0.092	0.102

In this second simulation test, no constraints were applied to the output variables of the unit to allow them to move freely.

However, constraints were applied to the control inputs and rate of input increments of the MPC local controllers. The constraints were applied due to the physical limitations of the actuators and to prevent damage of the dosing pumps. The flow rates constraints were set in range of 0 and +1 L/s for the sudfloc 3835 (q_a) and ferric chloride (q_b), and 0 and +2 L/s for hydrated lime (q_c). The rates of the input increments constraints (Δq_a, Δq_b, Δq_c) were set in the range of 0 and +5 L/s^2. Fig. 8 depicts the performance of the proposed FWMMPC compared with SMMPC and WMMPC schemes. Fig. 9 shows the corresponding control inputs applied to the MMPC schemes. From the RMSE values in Table 4 and visual inspection of Fig. 8, it can be deduced that FWMMPC scheme tracks the set points better than SMMPC and WMMPC schemes.

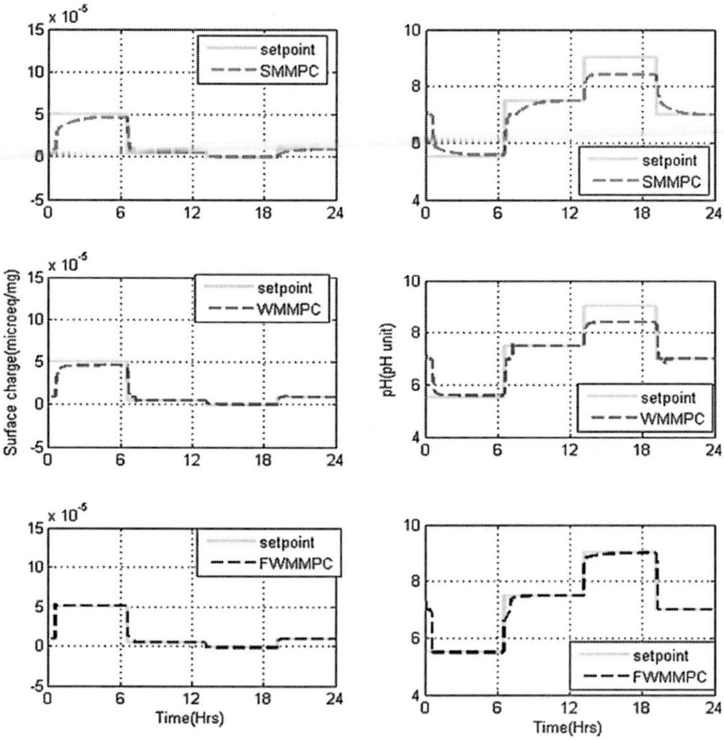

Figure 8: Performance of MMPC schemes with constrained MPC local controllers.

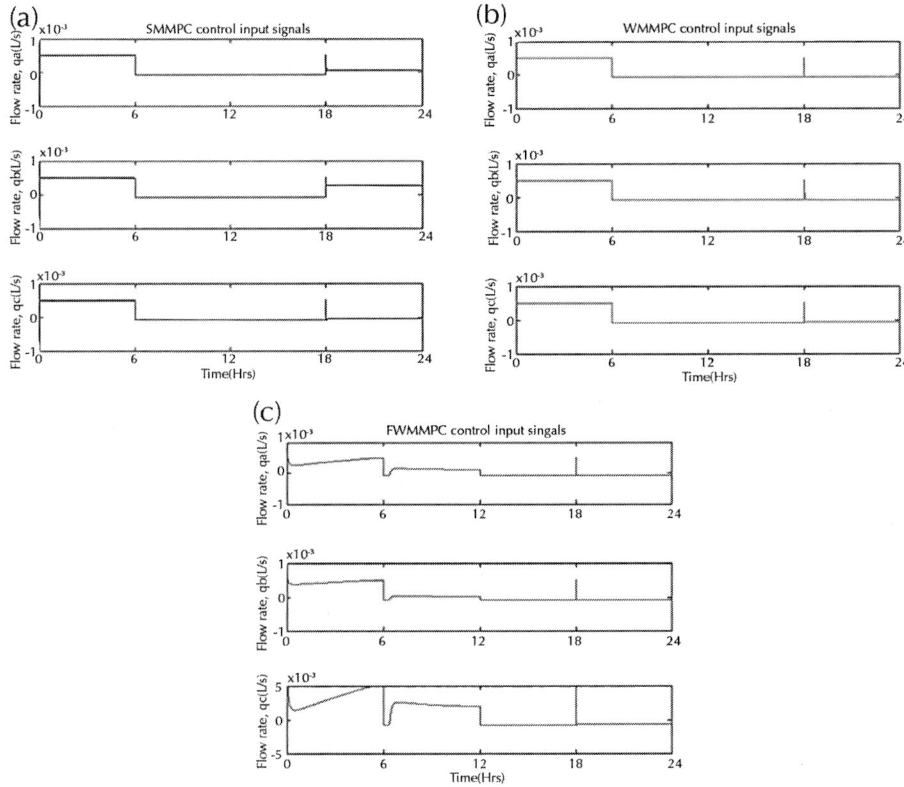

Figure 9: Control input moves of the MMPC schemes with constrained MPC local controllers.

The last simulation study investigates the performance of FWMMPC to reject disturbances in form sudden changes in water quality parameters, system faults and seasonal changes. Unmeasured and measured disturbances modelled in form of step changes in the flow rates of ferric chloride and hydrated lime were introduced to the dosing unit respectively. Fig. 10 depicts the disturbance rejection performance of the FWMMPC compared with the SMMPC and WMMPC schemes. Fig. 11 shows the control input moves to the control system. It can be observed that all the controllers perform satisfactorily. However, from the RMSE for the SMMPC, WMMPC and FWMMPC schemes shown in Table 4. FWMMPC has the smallest RMSE values. This indicates that the proposed control scheme has the best performance among the three controllers evaluated.

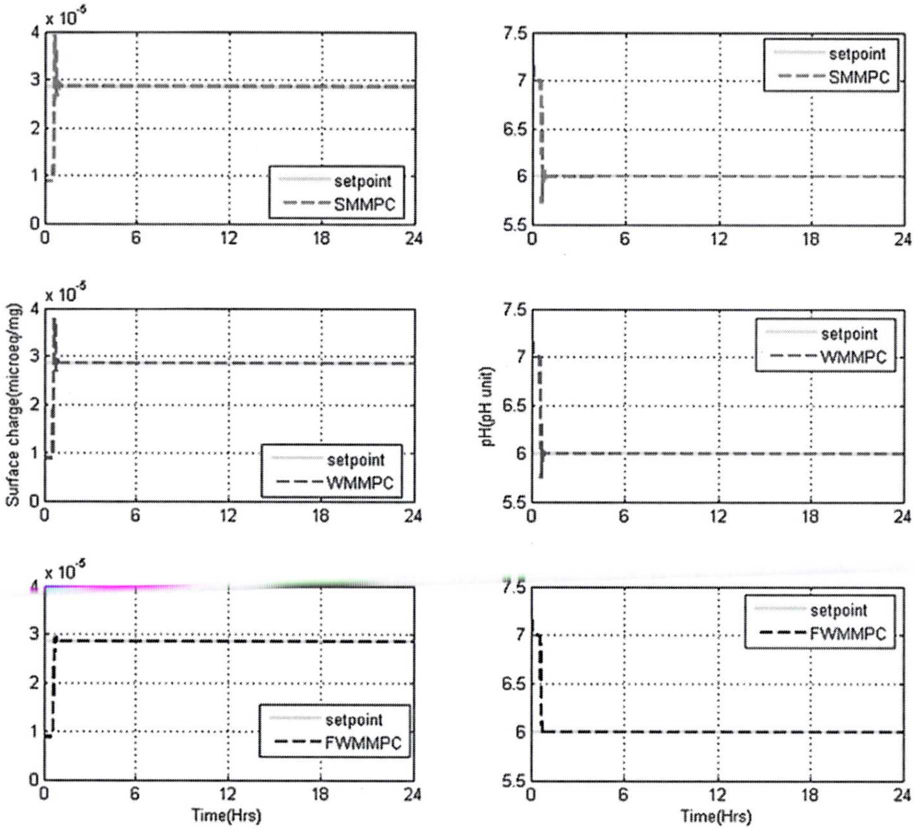

Figure 10: Disturbance rejection performance of MMPC schemes.

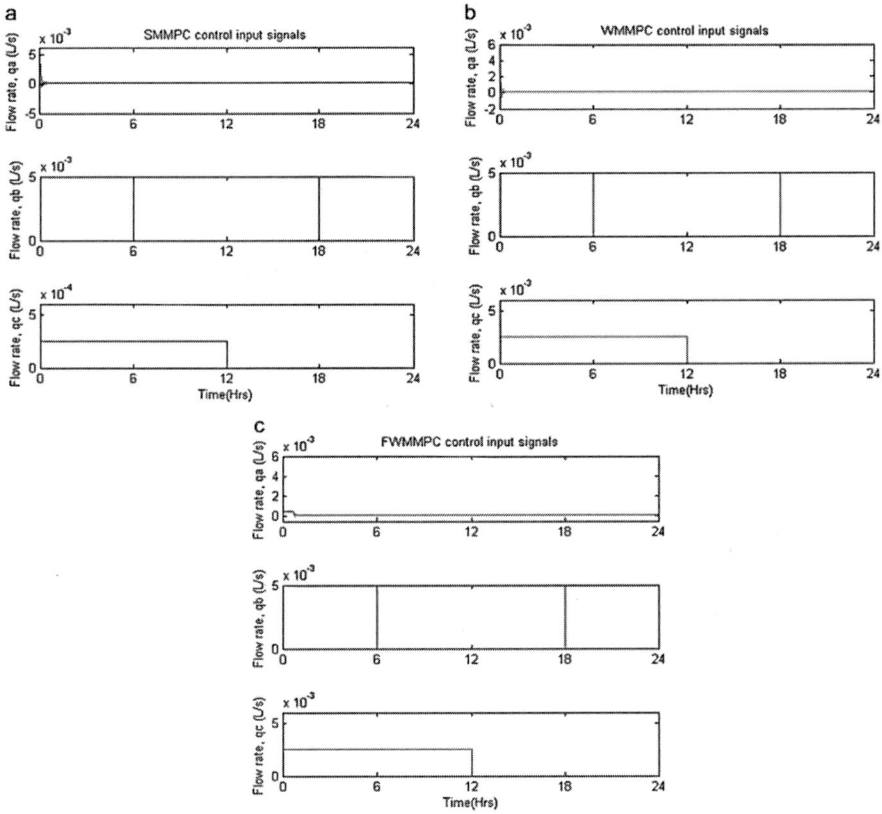

Figure 11: Control input moves of MMPC schemes.

CONCLUSIONS

This study has examined a multiple model predictive control based on fuzzy weighting approach to control the coagulation chemical dosing unit of water treatment plant over its entire range of operating conditions. For each model in the three operating conditions, a corresponding MPC local controller is designed to meet the control objectives. Simulation results show that the proposed fuzzy weighting multiple model predictive control performs better than SMMPC and WMMPC schemes under the same simulation scenarios. The proposed scheme will thus be suitable for control and optimisation of coagulation process in water treatment operations. Future works will include application of the proposed to scheme to a laboratory scale water treatment plant and evaluation of its performance.

ACKNOWLEDGMENTS

The support of Tshwane University of Technology, Pretoria for the work is gratefully acknowledged. The authors wish to thank Prof. J.T. Agee, management and staff of City of Tshwane and Rietvlei water treatment plant for their contributions and cooperation during the course of the study. The anonymous reviewers' comments on this paper are appreciated.

REFERENCES

1. A. Adgar, C.S. Cox, C.A. Jones, Enhancement of coagulation control using the streaming current detector, Bioprocess Biosystem Engineering (2005) 349–357.
2. O. Bello, Y. Hamam, K. Djouani, Dynamic modelling and system identification of a coagulant dosage system for water treatment plants, in: 3rd International Conference on Systems and Control, Algiers, 2013.
3. A. Bemporad, A. Morari, M. Morari, N.L. Ricker, Model Predictive Control Toolbox: User's Guide, The MathsWorks, Natick, 2013.
4. V.P. Evangelou, Environmental Soil and Water Chemistry: Principles and Applications, John Wiley, New York, 1998.
5. F. Di Palma, L. Magni, A multi-model structure for model predictive control, Annual Reviews in Control 28 (2004) 47–52.
6. A.W. Hermansson, S. Syafie, S.B. Mohd Noor, Multiple Model Predictive Control of nonlinear pH neutralisation system, in:

Proceedings of the IEEE International Conference on Industrial Engineering and Engineering Management, Macau, China, 2010, pp. 301–330.

7. K. Hu, J. Yuan, Multi-model predictive control method for nuclear steam generator water level, Energy Conversion and Management 49 (2008) 1167–1174.

8. B. Hua, A fuzzy logic based coagulant real time control scheme for water purification system, in: Proceedings of the IEEE International Conference on Mechatronics and Automation, Changchun, 2009.

9. M. Lawrynczuk,Modelling and nonlinear predictive control of a yeast fermentation biochemical reactor using neural networks, Chemical Engineering Journal 145 (2008) 290–307.

10. A.H. Mazinan, N. Sadati, Fuzzy multiple model predictive control of tubular heat exchanger, in: Proceeding of the IEEE World Congress on Computational Intelligence, Hong Kong, 2008, pp. 1845–1852.

11. K.S. Narendra, J. Balakrishnan, Improving transient response of adaptive control systems using multiple models and switching, IEEE Transactions on Automatic Control 39 (1994) 1861–1866.

12. NSF International, NSF Product and Service Listings, 2013 (Online). Available from: http://info.nsf.org/CertifiedChemicals/Listings.asp (accessed 19.08.13).

13. L. Ozkan, M.V. Kothare, C. Georgakis, Control of a solution copolymerization reactor using multi-model predictive control, Chemical Engineering Science 58 (2003) 1207–1221.

14. P.S. Paz, P.V. Ocampo, Automation and control of a water treatment plant process, in: Proceedings of 4th Mercosur Congress on Process System Engineering, 2009, p. 2009.

15. C.R. Porfirio, E. Almeida Neto, D. Odloak, Multi-model predictive control of an industrial C3/C4 splitter, Control Engineering Practice 11 (2003) 765–779.

16. L. Pourkarimi, M.A. Yaghoobi, M. Mashinchi, Efficient curve fitting: an application of multiobjective programming, Applied

Mathematical Modelling 35 (2011) 346–365.

17. J.S. Roger-Jang, N. Gulley,MATLAB: Fuzzy Logic Toolbox User's Guide, TheMathWorks, Inc., Natick, MA, 1997.
18. S. Roychowdhury, Fuzzy curve fitting using least square principles, Proceedings of the IEEE International Conference on Systems, Man, and Cybernetics 4 (1998) 4022–4027.
19. A. Shamasaddinlor, A. Fatehi, A.K. Sedigh, M.M. Karimi, Study of Multiple predictive control on a pH neutralisation plant, in: Proceedings of the 9th Asian Control Conference, Istanbul, 2013.
20. C. Yu, R.J. Roy, H. Kaufman, B.W. Bequette, Multiple-model adaptive predictive control of mean arterial pressure and cardiac output, IEEE Transactions on Biomedical Engineering 39 (1992) 765–778.
21. J. Zhang, J. Morris, Nonlinear model predictive control based on multiple local linear models, in: Proceedings of the American Control Conference, Arlington, VA, 2001, pp. 3503–3508.

Citations

CHAPTER 1

Yi Jing Chan, Mei Fong Chong, Chung Lim Law, D.G. Hassell, A review on anaerobic–aerobic treatment of industrial and municipal wastewater, Chemical Engineering Journal, Volume 155, Issues 1–2, 1 December 2009, Pages 1-18, ISSN 1385-8947, http://dx.doi.org/10.1016/j.cej.2009.06.041.

CHAPTER 2

Tiju Thomas, Nagaraju Kottam, Combining "chimie douce" and green principles for the developing world: Improving industrial viability of photocatalytic water remediation, Chemical Engineering Science, Volume 102, 11 October 2013, Pages 283-288, ISSN 0009-2509, http://dx.doi.org/10.1016/j.ces.2013.08.004.

CHAPTER 3

Tânia F.C.V. Silva, M. Elisabete F. Silva, A. Cristina Cunha-Queda, Amélia Fonseca, Isabel Saraiva, M.A. Sousa, C. Gonçalves, M.F. Alpendurada, Rui A.R. Boaventura, Vítor J.P. Vilar, Multistage treatment system for raw leachate from sanitary landfill combining biological nitrification–denitrification/solar photo-Fenton/ biological processes, at a scale close to industrial – Biodegradability enhancement and evolution profile of trace pollutants, Water Research, Volume 47, Issue 16, 15 October 2013, Pages 6167-6186, ISSN 0043-1354, http://dx.doi.org/10.1016/j.watres.2013.07.036.

CHAPTER 4

Ji í Jaromír Klemeš, Industrial water recycle/reuse, Current Opinion in Chemical Engineering, Volume 1, Issue 3, August 2012, Pages 238-245, ISSN 2211-3398, http://dx.doi.org/10.1016/j.coche.2012.03.010.

CHAPTER 5

Shirin Falamarzian, Omid Tavakoli, Reza Zarghami, Mohammad Ali Faramarzi, Catalytic hydrothermal treatment of pharmaceutical wastewater using sub- and supercritical water reactions, The Journal of Supercritical Fluids, Volume 95, November 2014, Pages 265-272, ISSN 0896-8446, http://dx.doi.org/10.1016/j.supflu.2014.07.017.

CHAPTER 6

S. Talvy, F. Debaste, L. Martinelli, E. Chauveheid, B. Haut, Development of a tool, using CFD, for the assessment of the disinfection process by ozonation in industrial scale drinking water treatment plants, Chemical Engineering Science, Volume 66, Issue 14, 15 July 2011, Pages 3185-3194, ISSN 0009-2509, http://dx.doi.org/10.1016/j.ces.2011.02.039.

CHAPTER 7

Evgenia Iakovleva, Ermei Mäkilä, Jarno Salonen, Maciej Sitarz, Mika Sillanpää, Industrial products and wastes as adsorbents for sulphate and chloride removal from synthetic alkaline solution and mine process water, Chemical Engineering Journal, Volume 259, 1 January 2015, Pages 364-371, ISSN 1385-8947, http://dx.doi.org/10.1016/j.cej.2014.07.091.

CHAPTER 8

Oladipupo Bello, Yskandar Hamam, Karim Djouani, Control of a coagulation chemical dosing unit for water treatment plants using MMPC based on fuzzy weighting, Journal of Water Process Engineering, Volume 4, December 2014, Pages 34-46, ISSN 2214-7144, http://dx.doi.org/10.1016/j.jwpe.2014.08.007.

Index

E

F

H

L

M

N

O

P

Q

R

S